Physiological Determinants of Exercise Tolerance in Humans

Physiological Determinants of Exercise Tolerance in Humans

Editors

B.J. Whipp

A.J. Sargeant

Portland Press

Published by Portland Press Ltd., 59 Portland Place,
London WIN 3AJ, U.K.
on behalf of the Physiological Society

In North America orders should be sent to Princeton University Press,
40 William Street, Princeton, NJ 08540, U.S.A.

ISBN 1 85578 026 7   ISSN 0969-8116

British Library Cataloguing in Publication Data
A catalogue record for this book is available from
the British Library

Cover illustration by T. Beer
Book design by A. Moyes

Typeset by Portland Press Ltd.
and printed in Great Britain by Information Press Ltd, Oxford, U.K.

Contents

Preface vii

Abbreviations ix

1
Muscle fatigue during high-intensity exercise 1
D.A. Jones

2
Neuromuscular determinants of human performance 13
A.J. Sargeant

3
Muscle energetics: aerobic strategies 29
D.L. Turner and H. Hoppeler

4
Skeletal muscle metabolism during high-intensity exercise in humans 49
M.E. Nevill and P.L. Greenhaff

5
The balance of carbohydrate and lipid use during sustained exercise: the crossover concept 61
G.A. Brooks

6
Water and electrolyte loss and replacement in the exercising human 77
R.J. Maughan

7
Cardiovascular function and oxygen delivery during exercise 93
N.H. Secher

8
Determinants and limitations of pulmonary gas exchange during exercise 115
S.A. Ward

9
The performance of the pulmonary system during exercise in athletes 135
B.J. Whipp

10
The effects of aging on exercise capacity 153
E.J. Bassey

Glossary 169

Index 173

Preface

The maxim of Joseph Barcroft (1872–1947) that: *"Exercise is not a mere variant of rest ... it is the essence of the machine"* succinctly characterizes the philosophy upon which this monograph is based. But what are the nature and essential control features of this 'essence'? The question imposes challenging demands in terms of human integrative physiology, requiring perspectives which range from the role of the smallest of all atoms, hydrogen (whose component electron and proton have important roles in energy transduction and fatigue), to the control mechanisms that mediate the complex interaction of the various physiological systems. An integrative approach, we believe, is therefore necessary both to provide an understanding of the normal determinants and limitations of human exercise performance and also to provide an appropriate frame of reference within which dysfunction can be interpreted. The issue should consequently be considered not only with respect to understanding the highest levels of athletic or sporting performance, but rather across the wide continuum of function including, for example, concerns such as mobility in the elderly, where it has important implications for the capability for independent living and hence quality of life.

The Physiological Society instituted a series of Teaching Symposia in an attempt to place recent physiological developments in appropriate conceptual context and hence provide a current 'state of the art'. These Symposia were designed to present frontier thinking on the topic in a form general enough for students beginning their study, yet detailed enough for students who are starting to specialize. This monograph is the result of two such Teaching Symposia, which considered both the determinants and limitations of muscular function during exercise and the systems which support the muscular energy transfer. It attempts to provide the material in a form which is not only logically sequential, but meets the dual goals of comprehensibility and rigour. Its focus is largely restricted to normal determinants and limits of human exercise tolerance. The influence of impaired systemic function is not at issue in this volume; pathophysiology is only considered when it helps explain features and consequences of system limitation.

The first four chapters address skeletal muscle. D.A. Jones introduces the thematic core of the monograph: muscle fatigue. This is followed by neuromuscular, energetic and metabolic considerations by A.J. Sargeant, D.L. Turner and H. Hoppeler, and M.E. Nevill and P.L. Greenhaff respectively. The role of substrate control and water and electrolyte balance is presented by G.A. Brooks, and R.J. Maughan. The focus then shifts to cardiovascular and respiratory concerns with chapters by N.H. Secher, S. A. Ward, and B.J. Whipp. And finally, the important issue of the effects of aging on performance is covered by E.J. Bassey. No attempt has been made to provide a single viewpoint or judgement on currently contentious issues. Rather, different perspectives are provided in different chapters in such cases: this, of course, more properly reflects the current 'state of the art'.

We wish to thank not only the authors who delivered their manuscripts promptly at, or near, our deadline but also the others who responded to our promptings, insistences and even threats without the slightest rancour. A special word of thanks must go to our secretaries Mrs Pat Chapman and Mrs Esther Verboom, and Ms Sarah Harrison of Portland Press, for their enormous help in bringing the volume to completion.

B. J. Whipp, London and A. J. Sargeant, Manchester, 1999

Abbreviations

acetyl-CoA	acetyl coenzyme A
acyl-CoA	acyl coenzyme A
ADH	antidiuretic hormone
ANF	atrial natriuretic factor
BTPS	body temperature and pressure; saturated with water vapour
C_{aO_2}	oxygen content of arterial blood
C_{vO_2}	oxygen content of mixed venous blood
CK	creatine kinase
C_{MvO_2}	oxygen content of muscle venous blood
CPT I	carnitine palmitoyltransferase
Cr	creatine
dm	dry mass muscle
EC	energy charge
FABP	free fatty acid binding protein
θ_F	fatigue threshold
F_{ACO_2}	fractional concentration of alveolar carbon dioxide
F_EO_2	fraction of average expired oxygen
FFA	free fatty acid
F_IO_2	fraction of average inspired oxygen
F–V	flow–volume
GABA	γ-aminobutyric acid
HGP	hepatic glucose production
IMP	inosine monophosphate
θ_L	lactate threshold
LPL	lipoprotein lipase
M-CK	myosin creatine kinase
MEFV	maximum expiratory flow–volume
Mi-CK	mitochondrial creatine kinase
MRS	magnetic resonance spectroscopy
MVC	maximum voluntary contraction
MVV	maximal volitional ventilation

NK	neurokinin
NMJ	neuromuscular junction
NMRS	nuclear magnetic resonance spectroscopy
O_2Def	oxygen deficit
P_a	arterial partial pressure
P_A	alveolar partial pressure
PCr	phosphocreatine
PFK	phosphofructokinase
P_i	inorganic phosphate
$\dot{Q}_{MO_2}$	muscle oxygen consumption
$\dot{Q}_T$	cardiac output
R	respiratory exchange ratio
Ra	rate of appearance
Rd	rate of disappearance
RQ	respiratory quotient
SNS	sympathetic nervous system
STPD	standard temperature and pressure; dry
TAG	triacylglycerol
TCA	tricarboxylic acid
TCAI	tricarboxylic acid intermediates
T_{tr}	vascular transit time
$\dot{V}_{CO_2}$	carbon dioxide output
$\dot{V}_D$	deadspace ventilation
V_E	ventilation rate
V_{max}	maximum velocity of shortening
$\dot{V}_{mito,max}$	maximal rate of mitochondrial respriation
$\dot{V}_{O_2}$	oxygen consumption
$\dot{V}_{O_2max}$	maximum oxygen utilization
V_{opt}	optimum velocity of shortening
V_T	tidal volume
V_{TO_2max}	whole-body $\dot{V}_{O_2max}$
$\dot{W}$	work rate
WBGT	wet bulb globe temperature

Muscle fatigue during high-intensity exercise

D.A. Jones

School of Sport and Exercise Sciences, The University of Birmingham, Birmingham B15 2TT, U.K.

Introduction

The word 'fatigue' is used to describe a wide variety of signs and symptoms related to the failure to sustain some form of physical activity and, as such, it is a common experience of everyday life which limits our work and play. Patients may suffer from premature or excessive fatigue as a result of respiratory or cardiovascular disease, or because of muscle wasting or due to specific defects in glycolytic or oxidative metabolism; at the other end of the spectrum, athletes are always anxious to extend the limits of their performance, which, in many cases, means delaying the moment when fatigue causes them to slow down.

Most reviews of fatigue begin with a definition of the word, such as fatigue being "an inability of a muscle or group of muscles to sustain the required or expected force" [1]. The difficulty with definitions such as this is that during the course of activity there are many changes in neuromuscular function, some of which lead to a reduction in performance, while other changes, such as an increase in temperature, can either enhance, or at least preserve, function. The extent of fatigue may appear greater for voluntary contractions than for tetanic stimulation, or may differ according to whether the muscle is tested at one frequency of stimulation compared with another, or if the muscle is allowed to shorten rather than being held isometric. It is important, therefore, in each situation, to specify the type of change in muscle function that is being described as fatigue.

The type of fatigue that will be dealt with in this review occurs as a result of high-intensity, relatively short-duration, exercise. It

may be useful at this stage to give an example of fatigue that has some relevance to normal activity, since most of the investigations described in the following sections have used very circumscribed and artificial forms of exercise. Fig. 1 shows the power output of a highly motivated athlete sprinting on a free-running treadmill. After an initial period of acceleration the power output falls away, so that by 30 s the output is only a little above half the peak power [2]. There are two types of question we can ask about this form of fatigue. The first concerns the site of failure: does the loss of performance arise because of a malfunction in, for instance, the neural pathways, at the neuromuscular junction (NMJ), or in any of the various processes concerned with the generation of force within the muscle fibre? The second question concerns the reason for the failure: if the failure is within the muscle fibre, is it due to a

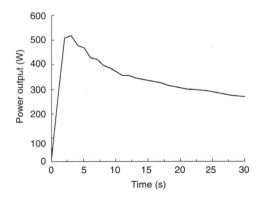

Fig. 1. Fatigue during sprinting
Power output of an experienced sprinter on a free running treadmill. Based on data from [2].

low concentration of ATP, high phosphate, a change in pH or other cause?

Central and peripheral factors

Muscular exercise involves the contractile machinery of the muscle under the control of the central nervous system; a failure in either would lead to loss of function and fatigue. Intense exercise gives rise to many sensations, such as breathlessness and pains in muscles and joints, which are generally unpleasant and might lead to a loss of central drive. The idea that muscular activity might be limited by the central nervous system is deeply rooted in sports and lay thinking. Even for fresh, unfatigued muscle it has been widely thought that voluntary effort is incapable of fully activating skeletal muscle — possibly, it is suggested, to prevent damage to tendons and bones. The test of this proposition is to compare the maximum voluntary contraction (MVC) force with that obtained by electrical stimulation of the muscle and, in 1954, it was clearly demonstrated that for the adductor pollicis muscle of the hand, supramaximal stimulation of the ulnar nerve generated tetanic forces that were comparable with the MVC [3]. An alternative for large muscle groups is to activate the motor nerve

branches through the skin with moistened pad electrodes [4].

Superimposing electrical stimulation has been used to test the extent to which a variety of muscle groups can be activated by a voluntary effort [5,6] (Fig. 2). The general conclusion from this type of investigation is that in a fresh, 'unfatigued' state most normal subjects, without any specific training, can activate the major muscle groups used in everyday life to within a few percent of their maximum.

▶ Failure of either the contractile machinery of muscle or the central nervous system leads to loss of muscle function and fatigue.
▶ Most normal subjects can almost fully activate their major muscle groups by voluntary effort in the unfatigued state.

Central fatigue

In 1954, Merton was the first to use electrical stimulation to study the question of central activation during fatiguing activity and concluded that full activation could be maintained during an MVC sustained for 3 min [3]. This has been confirmed with well-motivated and experienced subjects making contractions of the adductor pollicis and for prolonged contractions of the quadriceps lasting for 30–60 s [1], although for muscles of the lower leg there can be an appreciable component of central fatigue during voluntary contractions lasting about 1 min [7].

Most of the work on fatigue has concentrated on isometric contractions, but it is evident that whole-body exercise is a far more complex task than simply contracting one muscle group. An effective movement requires the co-ordinated contraction and relaxation of many muscles all over the body, and it remains a considerable challenge to assess the muscles' function while they are moving during complex activities. The limited work in this area will be discussed in the section on fatigue during dynamic contractions; for the moment we will concentrate on fatigue during isometric contractions.

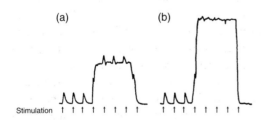

Fig. 2. Electrical stimulation superimposed on voluntary contractions
The muscle was stimulated with single twitches (arrows on baseline) while the subject made either a submaximal (a) or a maximal (b) contraction. Note that during the maximal contraction there is no additional force when the muscle is stimulated.

Peripheral fatigue

The neuromuscular junction

The idea that the NMJ may fail during fatigue is attractive, since exhausting the stores of neurotransmitter provides an obvious mechanism. In fact, failure may occur at a number of sites in and around the NMJ. Conduction may fail at the branch points of the fine terminal branches of the axons and there may be both pre- and post-synaptic mechanisms of failure. NMJ failure can be clearly demonstrated with isolated nerve–muscle preparations [8]. Where the number of acetylcholine quanta released is reduced, or the postsynaptic membrane is of a reduced sensitivity, as a result of poisoning (e.g. with curare) or disease (e.g. myasthenia gravis), the fresh muscle may be close to the threshold for successful transmission; however, for normal muscle, it is doubtful whether exercise places a significant stress on the NMJ.

To test NMJ function, it is necessary to compare force obtained by nerve stimulation with that obtained by direct stimulation of the muscle fibres, bypassing the NMJ. This can be done with isolated animal muscle preparations but it is difficult with human muscles *in situ*. Percutaneous stimulation with normally acceptable voltages activates the muscle only through the nerve endings. For this reason, most work on NMJ failure *in situ* has had to rely on less-direct experimental evidence: stimulating the motor nerve and recording the muscle action potential. This signal is known as the M wave and is the summation of individual fibre action potentials in the part of the muscle near the electrode; NMJ failure in a proportion of fibres should reduce the amplitude of the M wave. Merton used this technique to study fatigue in normal subjects and found that, whereas the force progressively declined during a sustained voluntary contraction, the muscle action potential remained constant, thus demonstrating the integrity of neuromuscular transmission [3]. In 1972, this conclusion was disputed by Stephens and Taylor, who presented conflicting evidence [9]. Since then, Bigland-Ritchie and colleagues have systematically examined the question of NMJ failure and the methodological problems

it entails, and have been unable to find any evidence of failure during maximal isometric contractions of up to 60 s [10].

On balance, it seems that NMJ failure is of little importance during short-duration, high-intensity activity but there may be unusual situations, such as prolonged electrical stimulation at high frequency or peripheral muscles working in cold conditions, where the NMJ is close to its limits.

Metabolic changes during fatigue

Fatiguing exercise entails large metabolic fluxes and changes in the concentration of muscle metabolites. It is understandable, therefore, that in the last 30 years a good deal of work has been directed to finding an explanation for fatigue in terms of altered metabolite levels affecting cross-bridge function, the main process using the chemical energy.

Much of the work on muscle energy metabolism in humans has made use of the needle-biopsy technique and analysis of the samples of muscle with linked enzyme assays (for example, see [11]). The technique requires samples to be taken and frozen rapidly but, despite the need for speed and the invasive nature of the procedure, it is a method that continues to be widely used and has provided a great deal of information about muscle metabolism. The main findings of the biopsy work have not been substantially altered by the introduction of magnetic resonance spectroscopy (MRS) applied to human muscle. The advantage of MRS is that the measurements are non-invasive and repeatable; the main disadvantage is that the technique is relatively insensitive, giving limited time resolution. MRS also entails a high investment in capital equipment and could never be used 'in the field' as the biopsy technique has frequently been.

The changes in muscle metabolites during a sustained maximal isometric contraction are shown in Fig. 3 for the human first dorsal interosseous, studied by MRS [12]. The concentration of phosphocreatine fell rapidly during the first 30 s and there was a concomitant rise in inorganic phosphate (P_i). Lactate con-

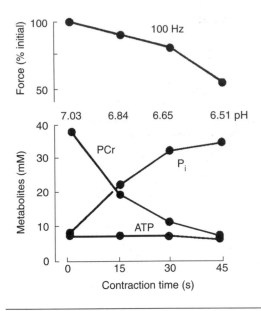

Fig. 3. Muscle metabolites during a sustained contraction of human muscle

MVCs were made under ischaemic conditions and the muscle tested with a brief 100 Hz tetanus at 15 s intervals. Metabolites were measured using MRS and the numbers above the metabolite data are the intracellular pH at those times [12]. Abbreviations used: PCr, phosphocreatine; P_i, inorganic phosphate.

centrations can be measured directly in biopsy studies but have to be inferred from the pH measurements when using MRS. The pH change seen in fatigued human muscle was about half a pH unit, from just over 7 to about 6.5. This corresponds to an accumulation of lactate that was approximately equimolar with the fall in phosphocreatine, whereas ATP concentration did not change significantly during the exercise. Although there are some disparities between results obtained by biopsy and MRS techniques, between animal and human preparations and between measurements made on whole muscle and single fibres, the main message is that ATP concentrations are remarkably well preserved in working muscle, and there is no simple relationship between deterioration of muscle performance and loss of ATP.

Metabolite concentrations and force production

The main steps in a normal cross-bridge cycle are shown in Fig. 4 with their probable relationship to the biochemical changes.

It can be seen that changes in the intracellular concentration of ATP, P_i or ADP might have profound effects on the cross-bridge cycle by altering the rates of transition between different states.

ATP

In the absence of ATP, cross-bridges remain in the attached state, which is the reason for the muscle stiffness after death (rigor mortis). Fatigued muscle relaxes slowly and in some circumstances resting tension will rise, and it is natural to wonder whether this signifies a low ATP level and the onset of rigor. However, experiments with skinned preparations [13] show that force generation is virtually independent of ATP in the range of concentrations that would be expected in normal and fatigued muscle. It would appear that the cross-bridge can function with an ATP concentration much lower than is likely to be found in any physiological situation except death.

Phosphate

P_i rises rapidly in fatiguing muscle and, as a reaction product of the actomyosin ATPase, might be expected to slow the release of P_i from the actomyosin complex, thereby preventing the transition of the cross-bridge from the low- to the high-force-generating state (see Fig. 4, step ii).

In skinned skeletal muscle preparations, P_i has been shown to have a marked effect on force generation [14], with most of the effect being seen with an increase of P_i from 0 mM to 10 mM. Human skeletal muscles, even at rest, have a relatively high P_i (6–10 mM, see Fig. 3), so it would seem that the force produced is less than optimum even in the fresh state. This might explain why, when there is an increase from about 8 to 20 mM within the first 15 s of contraction, there is comparatively little fall in isometric force [12] (Fig. 3).

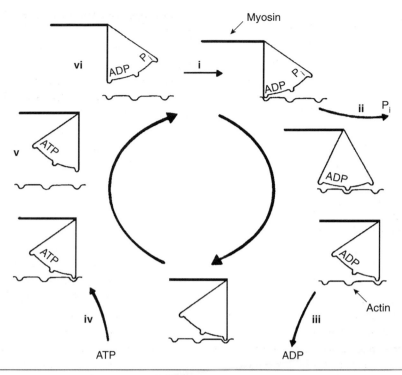

Fig. 4. Stages in the cross-bridge cycle showing the relationship between mechanical and biochemical steps

Attachment of myosin (which has bound to it the products of ATP hydrolysis) to actin (i) is a reversible process which will give stiffness to the muscle (i.e. it will resist if stretched) but will not itself generate force; this is sometimes referred to as the 'low-force' state. The release of P_i from the actomyosin complex (ii) is thought to initiate the changes which result in force generation with the cross-bridge moving into the 'high-force' state. Towards the end of the rotation phase of the myosin head, ADP is released (iii) and the actomyosin complex can then bind ATP (iv). Having done so, the actin and myosin dissociate with the ATP still bound to myosin (v). The ATP is hydrolysed and the products remain bound to the protein (vi); this last process is thought to activate the myosin head, making it ready to bind to actin again (i). ©1990, D.A. Jones.

It is possible that the discrepancy between the *in vivo* and *in vitro* preparations may be explained by differences in metabolite concentrations of different fibre types. Fast animal muscle has been shown to have P_i levels of 2–3 mM at rest, whereas slow muscles have resting concentrations of 8–12 mM [15], so that P_i accumulation may have a role to play in the fatigue of fast fibres but less so for slow fibres.

It has been suggested that loss of force in fatigue may be due to the combined actions of pH and P_i [16] rather than being due to the total phosphate. The pK of the reaction $HPO_4^{2-} +$ $H^+ \leftrightarrow H_2PO_4^-$ is close to 7, so the proportions of the mono- and di-basic forms of phosphate will vary greatly with change of pH in the physiological range. At rest, with an intracellular pH of just over 7, approximately two-thirds of the phosphate is in the dibasic form; however, as the muscle fatigues and pH falls, the proportion will change, so that by pH 6.5 approximately two-thirds will be in the monobasic form. The appearance of monobasic phosphate in the fibre will, therefore, lag behind that of the total phosphate and will correlate more closely with the change in force. In normal human muscle, the level of monobasic phosphate can rise to about 25 mM, the time course of which correlates fairly well with the extent of force loss [12,16]. However, where glycolysis is absent, as in patients with myophosphorylase deficiency, there is no change in pH and the monobasic phosphate will rise to only about 12 mM. Despite the much smaller rise in monobasic phosphate in these muscles, the force loss is as great, if not greater, than in normal muscle [12], demonstrating that monobasic phosphate cannot be the sole cause of force loss and suggesting that the relationships which have been reported are, most likely, coincidental rather than causal.

Intracellular pH

A decrease of intracellular pH as a result of glycolysis and H^+ accumulation can have a number of effects. High H^+ concentrations in skinned preparations lead to a reduction in the maximum force generated at saturating Ca^{2+} concentrations, directly affecting the ability of the cross-bridge to generate force. Another action is that low pH decreases the myofibrillar Ca^{2+} sensitivity, shifting the relationship between force and the intracellular calcium concentration to the right [17]. During maximal activation, this may be of little consequence as there seems to be an excess of Ca^{2+} released from the sarcoplasmic reticulum at high frequencies of stimulation. However, the reduced sensitivity to Ca^{2+} could lead to a lower force during partial activation at low frequencies of stimulation, such as used during the majority of activities.

Although a fall in pH may play some part in the development of fatigue, it cannot be the only cause. As discussed earlier in connection with the monobasic phosphate, fatigue develops even more rapidly than normal when glycolysis is absent and no rise in H^+ occurs, as with muscle poisoned by iodoacetate [18] or in patients with myophosphorylase deficiency [12]. In these muscles, pH rises slightly during the contraction as a result of phosphocreatine hydrolysis and the buffering action of P_i.

▶ NMJ failure is unlikely to cause fatigue during short-duration, high-intensity exercise; however, in some disease states, the NMJ may be close to its limits.

▶ Two main techniques have been used to analyse muscle energy metabolism in humans: (i) the needle-biopsy technique linked with enzyme assays; (ii) MRS.

▶ During sustained maximal isometric contraction, levels of phosphocreatine fall, P_i levels rise and lactate accumulates. ATP levels do not change significantly.

▶ In the cross-bridge cycle, release of P_i from the actomyosin complex is the force-generating event.

▶ The cross-bridge can function at very low levels of ATP.

▶ Loss of force may be due to changes in P_i and pH, either singly or together, but they are not the only or, perhaps, even the major, cause of fatigue.

Failure of activation

Some combination of increased phosphate and decreased pH almost certainly has a deleterious effect on force generation by the cross-bridge, but the extent to which it can explain fatigue, especially in normal, intact, human muscle remains unclear. Indeed, there are a number of experiments which suggest that the contractile elements retain almost normal function and that fatigue is primarily a problem of the processes of activation, i.e. the processes leading to the release of intracellular Ca^{2+}.

With isolated muscle preparations, high concentrations of caffeine lead directly to the release of Ca^{2+} from the sarcoplasmic reticulum, bypassing the normal excitation–contraction coupling mechanism. It is found that caffeine largely or completely reverses the decline in tension due to fatigue [19,20] (Fig. 5), showing that the cross-bridges are capable of generating normal forces. Measurements with intracellular Ca^{2+} indicators show that during the

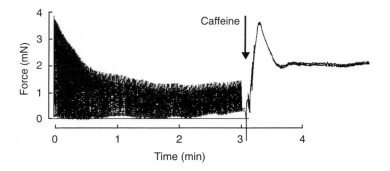

Fig. 5. The effect of caffeine on fatigued muscle

A bundle of fast mouse fibres was repetitively stimulated for 3 min after which time 30 mM caffeine was added and a contracture developed. Note that the peak of the caffeine contracture was of a very similar force to the first tetanus of the fresh muscle. Reproduced from [19] with permission. ©1989 The Physiological Society.

phase of rapid force loss in isolated mouse fibres there is a marked reduction in the internal free Ca^{2+} [21], which confirms the conclusion drawn from the caffeine experiments, that a major cause of force loss is a loss of activation.

> ▶ Caffeine can reverse the decline in tension during fatigue of isolated preparations.
> ▶ A major cause of fatigue is inactivation of the processes leading to release of intracellular Ca^{2+}.

Changes in excitability

One of the reasons for a decreased activation is the possibility that the action potential may not propagate along the surface and T-tubular membranes. There is ample evidence to suggest that the electrical properties of the muscle-fibre surface membrane can change during fatiguing activity with a reduction in the amplitude and slowing of the action potential waveform due to a slowing of conduction along the muscle fibre [22,23].

Stimulation of muscle at high frequency leads to a loss of force which, characteristically, recovers rapidly when the stimulation frequency is reduced. The increase in force on reducing the stimulation frequency is rapid, occurring within seconds, so it is unlikely that the recovery is associated with major metabolic changes in the muscle. It has been suggested that the changes in action potential size and form, and the loss of force seen during high-frequency stimulation, could be due to an excessive accumulation of K^+ or depletion of Na^+ in the T-tubules and extracellular spaces of working muscle [24]. There are large fluxes of K^+ from working muscles which appears in the venous drainage from working muscles, and blood levels can reach 5–6 mM during moderate exercise at around 100 W [25], while the concentration of K^+ in the interstitial spaces of the muscle would be appreciably higher than this. The bulk of the K^+ flux occurs through the T-tubular membranes; however, diffusion along the narrow T-tubules is relatively slow and, consequently, the concentration of K^+ in T-tubules at the centre of a muscle fibre is likely to be higher still. Increasing extracellular K^+ reduces the amplitude and prolongs the waveform of the resting action potential (Fig. 6), and mammalian muscle fibres, when placed in 15 mM K^+, become inexcitable [24]. Accumulation of K^+ in the extracellular space of a muscle would be expected to lead to a decrease in membrane potential; but reducing the resting membrane potential, by itself, would have the effect of making the muscle membrane more

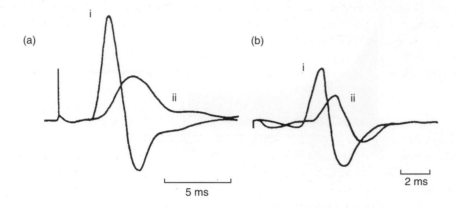

Fig. 6. Changes in the action potential waveform

(a) Human adductor pollicis, action potentials recorded after 1 s and 20 s stimulation at 50 Hz; (b) isolated mouse diaphragm. (i) Action potential of fresh muscle in medium containing 4 mM K^+; (ii) medium containing 15 mM K^+ [22].

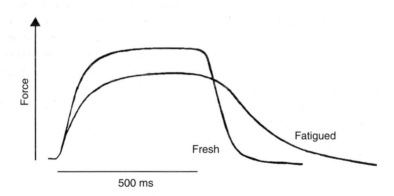

Fig. 7. Relaxation of force at the end of a brief tetanus of fresh muscle and after 30 s fatiguing contraction

Human first dorsal interosseous muscle of the hand. Reproduced from [27a] with permission. ©1989 The Physiological Society.

excitable by bringing the potential closer to the threshold at which the Na^+ channels open. There are, however, other changes which may result in a loss of excitability: it is known that when the membrane is mildly depolarized Na^+ channels become inactivated (slow inactivation), thereby reducing the inward Na^+ current, which could account for the loss of excitability [26].

Skeletal muscle is rich in ATP-dependent K^+ channels [27], which are closed in the presence of ATP and might be expected to open if the ATP concentration fell during fatigue. The role of these channels in skeletal muscle is unclear and it is by no means certain that they have a part to play in fatigue, since the channels appear to open only if the ATP concentrations should fall to around 0.1 mM, at least one order of magnitude lower than seen in exercising muscle.

> ▶ The electrical properties of the muscle-fibre surface membrane can change during fatigue, i.e. loss of amplitude and slowing of the action potential waveform.
> ▶ Changes in action potential may be due to high K^+ or low Na^+ levels in the T-tubules of working muscles.

Slowing of relaxation

Slowing of relaxation from an isometric contraction is characteristic of acutely fatigued muscle. The half-time of the exponential phase typically increases two- to three-fold (Fig. 7) and is electrically silent, so it is not the type of slow relaxation seen with myotonia (where the muscle fibre membrane fires repetitively). Under anaerobic conditions there is little or no recovery from this slowing but, when circulation is restored, recovery has a half-time of about 60 s, which is similar to the time course of phosphocreatine resynthesis. Although the simple exponential form of relaxation suggests a single underlying biochemical process, evidence is accumulating that there are two processes which can cause slowing: one is related to H^+ accumulation, the other is independent of pH change [27a].

The cause of slow relaxation has been debated for a number of years, there being two main possibilities. The first mechanism concerns Ca^{2+} re-accumulation by the sarcoplasmic reticulum, an ATP-dependent process which could be slowed in adverse metabolic circumstances. Using a Ca^{2+} indicator, Westerblad and Allen have shown that, for mouse muscle fibres, slowing of relaxation was not accompanied by a slowing of Ca^{2+} removal from the interior of the fibre [21]. This observation lends weight to the second hypothesis —

that the slowing of relaxation reflects a slowing of cross-bridge turnover, a change in function which has major implications for the force that can be sustained during dynamic contractions (see later).

> ▶ Acutely fatigued muscle relaxes slowly from isometric contraction.
> ▶ Slowing of relaxation is unlikely to be caused by slow re-accumulation of Ca^{2+} by the sarcoplasmic reticulum.
> ▶ Slowing of cross-bridge turnover is the most likely explanation for the slowing of relaxation.

Fatigue during dynamic exercise

In the first part of this chapter we have concentrated on the loss of force during high-force isometric contractions. In everyday life, however, most activities, such as walking or running, involve either shortening or lengthening of the muscles and consist of repetitive short contractions.

There is relatively little information about the effects of fatigue on force sustained during

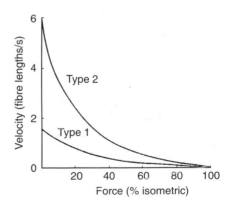

Fig. 8. Force–velocity relationships for different fibre types
Human type-1 (slow) and type-2 (fast) fibres, estimated from mixed preparations of human muscle obtained at surgery at 37°C.

movement (shortening or lengthening) as opposed to that generated during an isometric contraction. The loss of power illustrated in Fig. 1 is about twice as much as would be seen as a result of a sustained isometric contraction of the same duration [28]. The sprinting subject in Fig. 1 was a quality athlete, highly motivated and trained, so it is unlikely that central fatigue at a conscious level could have accounted for the more rapid loss of performance. Despite the potential importance of central fatigue during the movements involved in sporting activities, the phenomenon has not been systematically investigated, mainly because of the difficulty of stimulating and recording forces from muscles while on the move. However, two studies suggest that central fatigue is not a major problem during dynamic exercise. James et al. [29] looking at single-leg knee extension, and Beelen et al. [30] using sprint cycling on an isokinetic ergometer, found that electrical stimulation made very little difference to the forces generated by the quadriceps during high-intensity dynamic exercise lasting several minutes.

One reason why power output may show a larger fall than isometric force is because of a slowing of the fatigued muscle indicated by the slow relaxation from an isometric contraction of a fatigued muscle. Fig. 8 shows force–velocity curves estimated for fresh fast (type-2) and slow (type-1) human muscle fibres. Even if the two fibres generate the same isometric force, the force which can be sustained for a given velocity of shortening is much less for the slow type-1 fibre. If the muscle fibres were to become slower in their contractile properties during fatigue, the effect would be similar to a transformation of the fast fibres into slow in Fig. 8, a greater loss of the force during shortening as compared with the isometric contraction force.

There is growing evidence that the force–velocity characteristics of muscles do change with fatigue. There are reports of animal muscle studies where a decrease in the maximum velocity of shortening has been observed at a time when relaxation is slowing [31] and the observations of James et al. of a greater effect of fatigue on shortening as opposed to isometric

contractions in the quadriceps indicate that similar changes occur in human muscle [29].

As with the slowed relaxation from an isometric tetanus, there is likely to be more than one mechanism causing the slowing of shortening velocity. Acidification of a muscle fibre causes a reduction in the maximum velocity of unloaded shortening [32], but the precise step in the cross-bridge cycle where H^+ acts remains unknown. Either a reduction in ATP or an increase in ADP would decrease shortening velocity by reducing the rate of detachment of the cross-bridge at the end of the power stroke (see Fig. 4). However, the concentrations at which these changes are seen (less than 1 mM ATP or about 4 mM ADP; [13,14]) are outside the normal physiological range. It is most unlikely that in fatigued normal muscle ATP would fall below about 3–4 mM or ADP rise above 0.5 mM. Nevertheless, it is possible that there may be local concentration gradients within the fibre that could account for the slowing (for discussion, see [33]).

▶ During dynamic exercise, central fatigue is less of a problem than during isometric contraction.
▶ Force–velocity characteristics of muscle change with fatigue so the muscle becomes slower and power output declines more than isometric force.

Summary

Muscle contraction involves a chain of command from the higher centres, through the spinal cord, motor axons and the muscle fibre, and much of the discussion is about identifying the weak link in the chain that fails in fatigue. In a real metal chain there is no advantage in having one link stronger than the next, indeed it is wasteful to have anything other than uniform strength in all links. By analogy, it is likely that in the chain of command driving muscular activity, all the links will begin to fail at about the same time. One might expect, therefore, that the NMJ may be close to failure at the same time as conduction fails along the surface and T-tubular membranes, and Ca^{2+} release is impaired and the cross-bridges start to slow down. In evolutionary terms, it would be wasteful to have an NMJ or T-tubular system that would carry on working for ever, if the cross-bridges had long since ceased to function.

Bearing in mind the likelihood that many of the systems in the muscle will be running down at a similar rate, we return to the question asked at the start of this chapter concerning the reason for the loss of performance during the sprint illustrated in Fig. 1. Despite the sometimes conflicting and confusing evidence obtained from different experimental protocols and preparations, there is general agreement about the nature of the major changes which are taking place. These are summarized here.

> ▶ The loss of performance is largely due to changes in the muscle, i.e. it is of a peripheral rather than a central origin.
> ▶ In normal circumstances, the NMJ is unlikely to cause a loss of performance.
> ▶ The loss of power is caused by a change in the force–velocity characteristics of the fatigued muscle, of which there are two aspects: (i) a loss of isometric force, and (ii) a slowing of the shortening velocity of the muscle.
> ▶ A major cause of the loss of isometric force is a reduction in the amount of Ca^{2+} released with each action potential, i.e. a failure to activate the contractile elements.
> ▶ The reduction in shortening velocity is a result of a reduced rate of cross-bridge turnover.

These five points are a description of the changes occurring in acutely fatigued muscle and summarize the advances that have been made in this area of muscle physiology in the last 20 years. The controversy and uncertainties that remain are largely concerned with the causes of these changes. For instance, it is not clear why there should be less Ca^{2+} released as the muscle fatigues. Is there a failure of conduction down the T-tubules? Is there a change in Ca^{2+}-release mechanism? Does P_i or H^+ inhibit

the release? Likewise, why do cross-bridges turn over more slowly in fatigued muscle? Can we blame H^+, increased ADP levels or locally depleted levels of ATP? Answers to these questions will take the muscle physiologist to the level of molecular mechanisms concerning the interactions of metabolites with proteins.

References

1. Bigland-Ritchie, B., Jones, D.A., Hosking, G.P. and Edwards, R.H.T. (1978) Central and peripheral fatigue in sustained voluntary contractions of human quadriceps muscle. *Clin. Sci.* **54**, 609–614

2. Bobbis, L.H. (1987) Metabolic aspects of fatigue during sprinting. *In Exercise: Benefits, Limits and Adaptations* (Macleod, D., Maughan, R., Nimo, M., Reillt, T. and Williams, C., eds.), pp. 116–143, E. and F.N. Spon Ltd, London

3. Merton, P.A. (1954) Voluntary strength and fatigue. *J. Physiol.* **123**, 553–564

4. Edwards, R.H.T., Young, A., Hosking, G.P. and Jones, D.A. (1977) Human skeletal muscle function: description of tests and normal values. *Clin. Sci. Mol. Med.* **52**, 283–290

5. Rutherford, O.M., Jones, D.A. and Newham, D.J. (1986) Clinical and experimental application of the percutaneous twitch superimposition technique for the study of human muscle activation. *J. Neurol. Neurosurg. Psych.* **49**, 1288–1291

6. Gandevia, S.C., Allen, G.M. and McKenzie, D.K. (1995) Central fatigue. *Adv. Exp. Med. Biol.* **384**, 281–294

7. Kukulka, C.G., Russell, A.G. and Moore, M.A. (1986) Electrical and mechanical changes in human soleus muscle during sustained maximum isometric contractions. *Brain Res.* **326**, 47–54

8. Siech, G.C. and Prakash, Y.S. (1995) Fatigue at the neuromuscular junction. *Adv. Exp. Med. Biol.* **384**, 83–100

9. Stephens, J.A. and Taylor, A. (1972) Fatigue of voluntary muscle contractions in man. *J. Physiol.* **220**, 1–18

10. Bigland-Ritchie, B., Kukulka, C.G., Lippold, O.C.J. and Woods, J.J. (1982) The absence of neuromuscular function transmission failure in sustained maximal voluntary contractions. *J. Physiol.* **330**, 265–278

11. Chasiotis, D., Bergstrom, M. and Hultman, E. (1987) ATP utilization and force during intermittent and continuous muscle contractions. *J. Appl. Physiol.* **63**, 167–174

12. Cady, E.B., Jones, D.A., Lynn, J. and Newham, D.J. (1989) Changes in force and intracellular metabolites during fatigue of human skeletal muscle. *J. Physiol.* **418**, 311–325

13. Cooke, R. and Bialek, W. (1979) Contraction of glycerinated muscle fibres as a function of MgATP concentration. *Biophys. J.* **28**, 241–258

14. Cooke, R. and Pate, E. (1985) The effects of ADP and phosphate on the contraction of muscle fibres. *Biophys. J.* **48**, 789–798

15. Meyer, R.A., Brown, T.R. and Kushmerick, M.J. (1985) Phosphorus nuclear magnetic resonance of fast- and slow-twitch muscle. *Am. J. Physiol.* **248**, C279–C285

16. Miller, R.G., Boska, M.D., Moussavi, R.S., Carsonj, P.J. and Weiner, M.W. (1988) ^{31}P Nuclear magnetic resonance studies of high energy phosphates and pH in human muscle fatigue: comparison of aerobic and anaerobic exercise. *J. Clin. Invest.* **81**, 1190–1196

17. Allen, D.G., Westerblad, H. and Lannergren, J. (1995) The role of intracellular acidosis in muscle fatigue. *Adv. Exp. Med. Biol.* **384**, 57–68

18. Sahlin, K., Edstrom, L., Sjoholm, H. and Hultman, E. (1981) Effects of lactic acid accumulation and ATP decrease on muscle tension and relaxation. *Am. J. Physiol.* **240**, C121–C126

19. Jones, D.A. and Sacco, P. (1989) Failure of activation as the cause of fatigue in isolated mouse skeletal muscle. *J. Physiol.* **410**, 75P

20. Lannergren, J. and Westerblad, H. (1989) Maximum tension and force-velocity properties of fatigued, single *Xenopus* muscle fibres studied by caffeine and high K^+. *J. Physiol.* **409**, 473–490

21. Westerblad, H. and Allen, D.G. (1993) The contribution of $[Ca^{2+}]_i$ to the slowing of relaxation in fatigued single fibres from mouse skeletal muscle. *J. Physiol.* **468**, 729–740

22. Jones, D.A. and Bigland-Ritchie, B. (1986) Electrical and contractile changes in muscle fatigue. In *Biochemistry of Exercise VI* (Saltin, B., ed.), pp. 377–392, Human Kinetics Publishers, Champaign

23. Juel, C. (1986) Potassium and sodium shifts during *in vitro* isometric muscle contraction, and the time course of ion gradient recovery. *Pfluegers Arch.* **406**, 458–463

24. Jones, D.A. (1996) High and low frequency fatigue revisited. *Acta Physiol. Scand.* **156**, 265–270

25. Hallén, J. (1996) K^+ balance in humans during exercise. *Acta Physiol. Scand.* **156**, 279–286

26. Ruff, R.L. (1996) Sodium channel slow inactivation and the distribution of sodium channels on skeletal muscle fibres enable the performance properties of different skeletal muscle fibre types. *Acta Physiol. Scand.* **156**, 159–168

27. Spruce, A.E., Standen, N.B. and Stanfield, P.R. (1987) Studies of the unitary properties of adenosine-potassium channels of frog skeletal muscle. *J. Physiol.* **382**, 213–236

27a. Cady, E.B., Elshove, H., Jones, D.A. and Moll, A. (1989) The metabolic causes of slow relaxation in fatigued human skeletal muscle. *J. Physiol.* **418**, 327–337

28. Jones, D.A. (1993) How far can experiments in the laboratory explain the fatigue of athletes in the field? In *Neuromuscular Fatigue* (Sargeant, A.J. and Kernell, D., eds.), pp. 100–108, North Holland, Amsterdam

29. James, C., Sacco, P. and Jones, D.A. (1995) Loss of power during fatigue of human leg muscles. *J. Physiol.* **484**, 237–246

30. Beelen, A., Sargeant, A.J., Jones, D.A. and de Ruiter, C.J. (1995) Fatigue and recovery of voluntary and electrically elicited dynamic force in humans. *J. Physiol.* **484**, 227–235

31. De Haan, A., Jones, D.A. and Sargeant, A.J. (1989) Changes in velocity of shortening, power output and relaxation rate during fatigue of rat medial gastrocnemius muscle. *Pfluegers Arch.* **413**, 422–428

32. Chase, P.B. and Kushmerick, M.J. (1988) Effects of pH on contraction of rabbit fast and slow skeletal muscle fibres. *Biophys. J.* **53**, 935–946

33. Allen, D.G., Lannergren, J. and Westerblad, H. (1995) Muscle cell function during prolonged activity: cellular mechanisms of fatigue. *Exp. Physiol.* **80**, 497-527

2

Neuromuscular determinants of human performance

A.J. Sargeant[1]

Institute for Fundamental and Clinical Human Movement Sciences, Vrije University, Amsterdam, The Netherlands, and Neuromuscular Biology Research Group, Manchester Metropolitan University, Manchester, U.K.

Introduction

The role of skeletal muscle is to generate force. This chapter explores how the human (neuro)-muscular system is structured and organized to meet the need for both instantaneous and sustained mechanical output. Some of the intrinsic and extrinsic factors which can modulate the output are described, including reference to the innate 'plasticity' of the system. The structural/molecular basis of the force-generating process is not dealt with in detail since detailed accounts can be found in standard textbooks and reviews (e.g. [1–5]).

Muscle function has been studied using a wide variety of preparations. These range from sections of single fibres from which the cell membrane has been removed to whole muscles or groups of muscles acting about a joint in the intact animal. It should be noted here that in the intact human most studies of skeletal-muscle function necessarily involve a group of muscles acting about a joint (or joints). Each preparation has its merits and limitations. There is no universal 'ideal' preparation — rather the question to be addressed will determine the most appropriate one in each case, and sometimes a combination will be needed.

It may be useful in this context to remember that, from the point of view of neural control and patterns of muscle-fibre recruitment, the motor unit can be thought of as the final unit of the system. The motor unit consists of a motor neuron in the spinal cord and all of the muscle fibres that it innervates via its axonal branches. Since the pattern of use is a primary determinant of a muscle fibre's contractile and metabolic properties, it follows that the muscle fibres within a motor unit will be relatively homogenous. In relation to the contractile and metabolic properties, it is the muscle fibre or component units of the fibre (the sarcomeres) which are the ultimate focus of attention. Nevertheless, for a proper understanding of human physical performance, it is necessary to understand how the variability within the component units is organized to optimize performance in relation to the dimensions and structure of the human body acting within the physical environment.

Fundamental muscle properties

Notwithstanding the range of preparations studied, there are two fundamental relationships that determine the mechanical output of muscle. These are the length–tension relationship and the force–velocity relationship.

Length–tension relationship

It has been known since the 19th century that the isometric tension a muscle can generate depends on its length. The active tension is generally believed to be the consequence of the formation of cross-bridges in the region of overlap of actin and myosin filaments within each sarcomere. In addition, there is at long muscle lengths an increasing element of passive tension as structural elements both outside and within the muscle fibre are stretched. To obtain the active component of the length–tension relationship, the passive component is usually

[1]Address for correspondence: Neuromuscular Biology Research Group, Manchester Metropolitan University, Hassall Road, Alsager ST7 2HL, U.K.

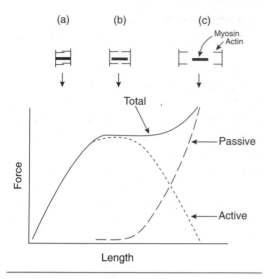

Fig. 1. Relationship between muscle length and the active (···) and passive (---) isometric tension

The amount of actin and myosin overlap which determines the active component is indicated for (a) a short length where actin filaments from opposite ends of the sarcomere overlap and force is reduced; (b) optimum length where the greatest active force is generated due to the maximum number of cross-bridges; and (c) at a long length at which there is no overlap and no cross-bridges are formed.

subtracted. The observed active length–tension relationship has been found to fit reasonably closely to that predicted from the degree of actin and myosin filament overlap in accordance with the cross-bridge theory (Fig. 1).

Force–velocity relationship

Not only does force vary according to the degree of myofilament overlap, as described by the length–tension relationship, but it also varies with the velocity of shortening or lengthening. The force–velocity relationship of muscle was first described in the 1920s and 1930s and the basic description remains much the same, as shown in Fig. 2. Thus, as velocity of shortening increases, the force that is generated falls in a hyperbolic fashion, eventually reaching zero at a velocity defined as the maximum

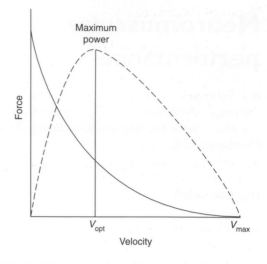

Fig. 2. Force–velocity and power–velocity relationship of skeletal muscle

Maximum power is generated at the optimum velocity (V_{opt}), which is about 30% of the maximum velocity of shortening (V_{max}) when force is zero.

velocity (V_{max}). During lengthening, the force generated by an active muscle in resisting that lengthening increases above that attained at zero velocity (isometric) before plateauing (not shown in Fig. 2, but see [5] for discussion and illustration of the phenomenon).

The force–velocity relationship reflects the structure of the cross-bridges and the kinetics of their attachment and detachment [5].

The mathematical product of force and velocity is power and this will have its own distinctive parabolic relationship with velocity, reaching a maximum at around an optimum (V_{opt}), which in isolated muscle preparations is around 30% of V_{max}.

▶ The isometric tension that a muscle can generate depends on its length.
▶ Active tension results from the formation of cross-bridges where actin and myosin filaments overlap in the sarcomere.
▶ The force generated by muscle is dependent on the velocity of shortening or lengthening of that muscle.

Types of contraction

The mechanical output required from skeletal muscle *in vivo* may require one of, or a combination of, three types of 'contraction', often referred to as isometric, eccentric or concentric contractions.

Isometric contraction

Isometric contraction implies that the muscle is activated but is held at a constant length. In fact, in whole-body human and other animal studies, it is usually defined operationally as an activation of muscle(s) when the joint(s) spanned by the muscle(s) are held in a fixed position. The important point here is that there will always be an element of shortening of the muscle due to (among other things) stretching of the tendinous connections and compression of joints. For example, in metabolic experiments involving short-duration, repetitive, isometric contractions (i.e. with a fixed joint position), the whole activation period may be taken up with shortening of the muscle fibres as tendons are stretched or joints compressed. In addition to this situation, there may be rather complex and simultaneous lengthening and shortening of sarcomeres in series along the length of the individual muscle fibres, especially when fatigue is occurring due to heterogeneity of metabolic and mechanical properties along the length of a fibre.

The generation of isometric force is, for largely technical reasons, the most commonly studied form of contraction. It is also of considerable importance, stabilizing joint complexes during locomotion and maintaining posture.

Eccentric contraction

Eccentric contractions occur when the muscle is activated but, instead of shortening, it is lengthened due to an external force which exceeds that generated by the degree of activation. This type of muscle use is a normal element of daily activity. Everyday examples are the lowering of a weight, or walking downstairs. In these circumstances work is done on the muscle and energy is absorbed. The muscle acts as a 'brake'.

Concentric contraction

Concentric contractions are those where the muscle is activated and shortens, power is generated and work is done by the muscle. It is of fundamental importance in all human physical activity. Largely for technical reasons, properties of human muscles have been studied much less often under dynamic than isometric conditions, often on the implicit assumption that isometric function will systematically reflect the functional status of the muscle for concentric contractions. It is now clear, however, that this is not the case, for reasons described in Chapter 1 in this volume.

The above nomenclature is open to criticism: how can there be a *contraction* when the length remains constant or increases? Nevertheless, the terms are commonly used in muscle physiology, and seem unlikely to disappear in the near future.

Stretch–shortening and other preactivation contractions:

As well as the pure forms of contraction, there are many situations where a combination of different types of contraction occur, the most important of which may be the stretch–shortening cycle where an eccentric contraction of the muscle immediately precedes a concentric, power-generating phase. For example, in running, leg extensor muscles generate power at each stride by concentric contraction, but in addition they also act as a 'brake' or 'shock absorber' on landing, absorbing some of the potential and kinetic energy generated in the previous push-off by the other leg. Some of that energy is stored in elastic structures and can contribute to the power output in the next push-off [6]. The re-use of stored elastic energy in this way is common throughout the animal kingdom, but perhaps the most well-known development of the principle in mammals is seen in the kangaroo; less expected is the contribution that it makes to the achievement of high, sustained, running speeds in the camel [7].

It is not necessary, however, for a muscle to be lengthened during a preactivation phase in order for energy to be stored in the series elastic elements of the muscle–tendon complex. Pre-activation can occur while the whole muscle–tendon complex remains at constant length or is only slightly lengthened. In this case, the contractile elements may remain isometric or even shorten, stretching and storing elastic energy in the tendinous elements. During the subsequent power-generating phase, this 'elastic energy' can contribute to the delivered force, and the magnitude of this contribution will increase as the speed of movement increases. This mode of functioning may be of particular significance in relation to bi-articular muscles which cross two joints and which, functioning in this way, may act to translocate power to more distal limb segments (for further discussion of how these effects modify the force–velocity characteristics of muscle–tendon complexes *in situ*, see Chapter 6 in [1]).

▶ In isometric contraction the joints spanned by the activated muscle are held in a fixed position.
▶ In eccentric contraction the activated muscle is lengthened due to an external force which exceeds the degree of activation.
▶ In concentric contraction the activated muscle is shortened, generating power, and work is done by the muscle.
▶ In many situations a combination of different types of contraction occur, e.g. stretch–shortening.

Human muscle fibre types

Consideration of how human muscle meets the demands for different types of mechanical output must take account of the variation in the contractile and metabolic properties of the muscle fibres of which the whole muscle is composed, and their pattern of recruitment. On the basis of histochemistry of serial sections, human muscle fibres are commonly divided into three major types: I, IIA and IIB. These types may be more-or-less analogous to muscle fib-res from animals, which have been classified on the basis of their directly determined functional properties as slow, fast (fatigue resistant) and fast (fatiguable), respectively (Table 1). It should be noted, however, that these functional properties have been determined less often for human muscle fibres and, in absolute terms, there are important differences between the analogous fibres from humans and other mammalian species. Furthermore, it should be realized that conventional histochemical techniques have been designed to differentiate muscle fibres into discrete types, and, although this categorization is convenient, it can be misleading since there is a continuum in most, if not all, metabolic and contractile properties among muscle fibres.

In recent years it has become possible to show that there is a complex set of genes controlling the expression of the contractile proteins and that these are closely linked to those determining metabolic properties. Of the contractile proteins, the isoform of the myosin heavy chain that is expressed seems to be the primary determinant of a muscle fibre's maximum velocity of shortening, and hence power. In human muscle there appear to be three main isoforms: type-I (slow); type-IIA (fast); and type-IIX (the fastest). Unfortunately, and confusingly, this last isoform (so designated because of its close similarity to the IIX isoform identified in other mammalian species) is associated with the human muscle fibre type previously designated as type-IIB on the basis of

Table 1. Principal systems of muscle fibre-type classification

Human muscle fibres	Contractile/ metabolic	Physiological
Type-I	Slow, oxidative	ST (slow, fatigue resistant)
Type-IIA	Fast, oxidative	FFR (fast, fatigue resistant)
Type-IIB (IIX)	Fast, glycolytic	FFS (fast, fatigue sensitive)

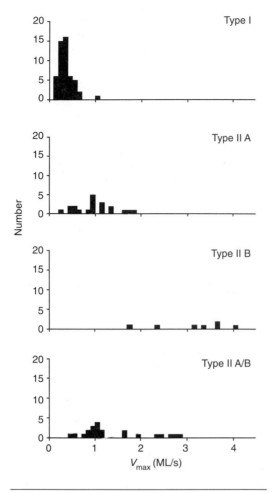

Fig. 3. Distributions of V_{max} values from human muscle for fibres characterized according to the myosin heavy-chain isoform expressed, i.e. types I, IIA, IIB (IIX) and fibres which express both IIA and IIB(X), designated IIA/B in this classification

Note the continuum of V_{max} values across the fibre groups, also that although the mean V_{max} for the type-II fibres is approx. four times that of the type-I fibres, there is a wide range of V_{max} values, indicating that the V_{max} of the fastest IIB fibres might be about eight times that of the type-I. Abbreviation used: ML, muscle length. Reproduced with permission from [9]. ©1993, The Physiological Society

conventional histochemistry (see [8] for a discussion of this point).

One of the more important observations is that, whereas type-I and type-II myosin heavy-chain isoforms are not usually found in the same fibre, the majority of human type-II fibres have both the IIA and the faster IIX isoforms in variable amounts [9–11]. Indeed, it is notable that in normal healthy humans there are very few fibres that express purely the IIX isoform. The relatively large proportions of what have previously been defined by conventional histochemistry as IIB fibres presumably included many 'hybrid' fibres which co-expressed significant amounts of the IIA isoform. The functional consequences of this pattern of contractile protein expression are indicated in Fig. 3, which shows the maximal velocity of shortening (V_{max}) of skinned human muscle fibres [9]. In this figure it can be seen that what the authors then referred to as the IIA/B fibres (IIA/X) are intermediate between those fibres which expressed only IIA or IIB (IIX) forms. Thus there is a continuum of V_{max} values but with considerable variation even within those fibres that expressed only one isoform. The latter observation suggests that even though the myosin heavy chain might be considered to be the primary 'molecular motor', it is modulated by other factors, such as the myosin light-chain composition — although at the time of writing this still has to be demonstrated in human muscle (see [12]).

As might be expected, there is a close link between the expression of the contractile proteins, the metabolic properties of the fibres and the concentration of substrates. Thus the maximally Ca^{2+}-activated myofibrillar ATPase activity is related to the myosin heavy-chain isoform(s) expressed in human fibres [13]. In turn, this is reflected in the resting concentration of phosphocreatine, which provides a temporal and spatial buffering system for ATP turnover. In human muscle fibres, as the intrinsic speed of cross-bridge cycling increases with the proportion of the IIX isoform expressed, so does the resting concentration of phosphocreatine, to meet (for example) the increased rate of depletion during maximal exercise (Fig. 4 and Fig. 5) [11].

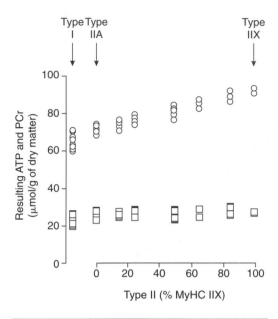

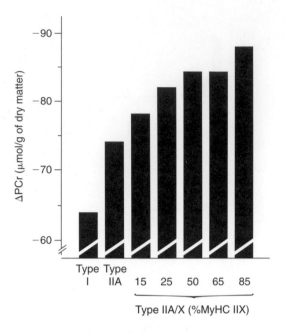

Fig. 4. ATP and phosphocreatine content in resting fibres characterized according to their myosin heavy-chain composition
Contents of ATP ($\square$) and phosphocreatine ($\circ$) are shown. Each value represents one individual pool of fibres. Hybrid IIA/X fibres were pooled according to the percentage of their myosin heavy-chain IIX component. Abbreviations used: PCr, phosphocreatine; MyHC, myosin heavy chain. Reproduced with permission from [11]. ©1996, The Physiological Society.

Fig. 5. Calculated change in phosphocreatine content (ΔPCr) in human muscle fibres after a maximal 25 s sprint on an isokinetic cycle ergometer
Single fibres were microdissected from a needle biopsy of the vastus lateralis muscle and the data corrected to allow for the resynthesis of PCr during the period between the end of exercise and freezing of the biopsy (~5 s). (See Fig. 4 for abbreviations; data taken from Table 1 in [11]).

▶ Human muscle types can be divided into three major types: type-I, type-IIA and type-IIB (IIX).
▶ The myosin heavy-chain isoform that is expressed is the primary determinant of muscle fibre type.

Chronic plasticity

In the human muscle fibre, type differentiation seems largely complete in humans a few years after birth, although the precise modulating role that training or other interventions play in relation to the predetermined genetic code is not clear. It has, however, been recognized for many years, since the classic cross-innervation studies of Buller and colleagues [14] and subse-

quent chronic stimulation studies of Salmons and Vrbová [15], that the pattern of contractile activity is an important factor determining the contractile and metabolic properties of muscle fibres (but not the sole factor, see e.g. [16]). The actual signals that trigger modification of the muscle fibres' molecular structure and function have not yet been fully elucidated, although mechanical events, energy status of the cell, or excitation/contraction coupling processes are all possible candidates. (For recent reviews of these major topics, see [3,17].)

Motor unit recruitment

Muscle fibres are organized into motor units consisting of one motoneuron and a variable

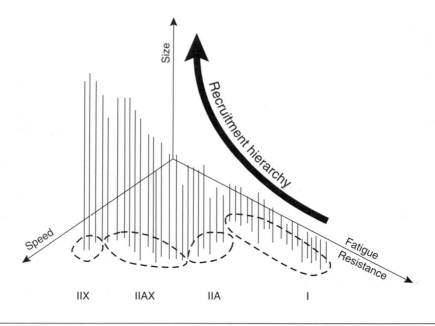

Fig. 6. Motor unit variability in size, intrinsic speed and fatigue resistance

The boundaries of the discrete fibre-type groups are given by the dashed lines but inreality there is a continuum of properties. The hierarchical pattern of successive recruitment of larger motor units is indicated by the arrow (see text for further discussion).

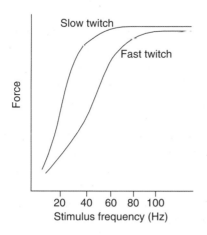

Fig. 7. Relationship between force and stimulation frequency for slow- and fast-twitch muscle fibres

Note this relationship is temperature dependent and shifts to the right as contraction velocity increases.

number of muscle fibres of the same type. The gradation of mechanical output from the muscle may be achieved by recruitment of an increasing number of units or by rate coding, i.e. by increasing the stimulation frequency.

The size principle of motor unit recruitment

In many but not all situations there appears to be an orderly hierarchy of motor unit recruitment based on size, as originally proposed by Henneman and Mendell [18]. Thus, at low levels of recruited force, small (low-force) motor units with thin axons and small, easily excitable motoneurons are recruited. As the requirement for force increases, there is a systematic recruitment of larger (high-force) motor units so that the force steps caused by successive recruitment remain rather constant when expressed as a proportion of the prevailing force output. Thus, not surprisingly, and as a consequence of the pattern and duration of activation implicit in such a recruitment pattern, there is in general an association with increasing size and increasing fatiguability and power output — this corresponds with a progression from type-I to -IIA to -IIX(B) (Fig. 6). Superimposed upon this, however, there is an important element of task-related recruitment of motor units and rate coding (for review see [19]).

Rate coding and fatigue

The relationship between the frequency of stimulation and the force generated is characterized by the force–frequency curve (Fig. 7), which reflects the degree of fusion attained. The relative importance of rate coding and motor unit recruitment in human muscles is not established with certainty and may vary between, for example, the small muscles of the hand and the large locomotory muscles. In cycling exercise involving large locomotory muscles, there is evidence to suggest that even at relatively low levels of force output (~50% of maximum dynamic force) all motor units have been metabolically active (Fig. 8) [21,23,23a]. This implies a significant element of rate coding such that the relative activation frequency might decrease the higher that a motor unit is in the recruitment hierarchy. In this way, fatigue-sensitive, high-power motor units might contribute to the mechanical demands of the exercise but the onset of fatigue may be delayed due to the relatively low level of activation. Nevertheless, it does seem that in high-intensity exercise the fatigue that occurs in terms of the whole-muscle power output is the result of a mismatch between activation level and fatiguability of fast motor units; thus the observed fatigue, assessed as a loss of maximum power, could be the consequence of a selective fatigue of a relatively small population of fast, fatigue-sensitive, motor units which in the fresh state can generate very high power output [11,22,23].

It will be appreciated that, from a control systems viewpoint, the greatest sensitivity will be in the steep part of the force–frequency curve, and indeed it is notable that in human exercise the firing frequency of motor units is usually rather low (10–30 Hz) — at which level the motor units will not achieve a fused tetanus.

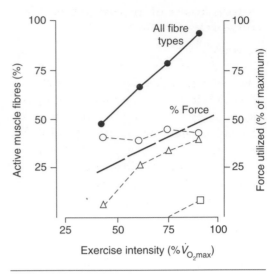

Fig. 8. Proportions of maximal force utilized (● — ●) and muscle fibres active (— —), in relation to the exercise intensity expressed as a percentage of maximum oxygen uptake (%$\dot{V}_{O_2,max}$) in human cycling exercise
Values are given for the total (● – – ●) and component fibre-type populations: type-I (○ – – ○); type-IIA (△ – – △); type-IIAX and -IIX combined (□ – – □). Reproduced with permission from [23a]. ©1985, The Physiological Society.

The full implications of this strategy in terms of the energy cost of contractions have yet to be explored. It should also be noted that the pattern of stimuli during contraction is not constant. Typically, the interval between the initial stimuli is much shorter, producing so-called 'doublets'. The effect of this is to enhance force development. (For further discussion, see Chapter 45 in [2]).

▶ The size of the motor units within muscle fibres correlate with the force requirement of that muscle.
▶ Force output may be controlled by increasing the number of motor units recruited and the frequency at which they are activated.
▶ Large, powerful, fatigue-sensitive motor units may be recruited at relatively low frequencies in order to delay the onset of fatigue.

Determinants of maximal strength and power in humans

Activation

Clearly the first prerequisite for generation of maximal force is that the muscle is fully activated. Whether or not a muscle is fully activated during a voluntary maximum effort can be tested by applying electrical stimulation to the muscle or nerve — if activation by the nervous system were incomplete, an increased force should be elicited. Despite many anecdotal accounts describing 'superhuman' efforts under extreme stress, the consensus of evidence — at least for the commonly used limb muscles — is that subjects can achieve maximal or near maximal (>95%) contractions voluntarily in most efforts performed under isometric conditions. Whether or not maximal activation of muscle can be achieved voluntarily in dynamic exercise has proven technically more difficult to determine, although the increasing

consensus of evidence seems to indicate that it can, at least in concentric contractions under fresh conditions. This conclusion is based on (i) studies where electrical stimulation has been superimposed on, or compared with, maximal voluntary dynamic contractions [23,24]; (ii) electromyography (EMG) evidence which although indirect does not immediately suggest a failure of neural drive in fatigue [20]; (iii) the fact that rather low maximal forces are generated in these fast dynamic contractions (+50% of maximum isometric force), which would seem to mitigate against the possibility of inhibitory reflexes influencing activation.

In the context of human performance, it should be noted that, when there is maximal activation of an agonist muscle in a voluntary contraction, the net torque generated round the joint will be reduced if there is a simultaneous activation of the antagonist muscles. Recent studies of the ankle dorsi- and plantar-flexors

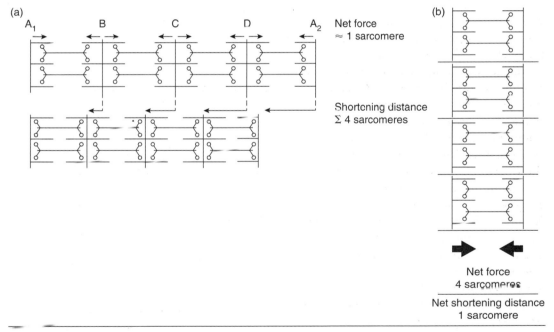

Fig. 9. Schematic illustration to show the force generated by four sarcomeres arranged (a) in series and (b) in parallel

(a) The forces either side of the Z lines B, C and D cancel one another out, thus the net force of the system is only the forces generated at A_1 and A_2, i.e. equivalent to one sarcomere. (b) The forces generated by all four sarcomeres add up to give four times the net force delivered by (a). [Note: conversely, the distance shortened (and hence velocity) will be four times greater in system (a) compared with system (b). Thus the maximum power (i.e. force $\times$ velocity) will be the same in both systems, although it will be achieved at different optimal velocities.]

indicate that this effect is most marked at extreme ankle angles where co-activation may reduce the net torque by ~20–30% [24a].

Isometric strength and muscle size

Assuming adequate activation, the determinant of the maximum strength must be intrinsic to the muscle. Cross-bridges act as independent force generators but they are arranged in sarcomere units with opposing forces, i.e. cancelling forces generated either side of the Z line. Thus it should not matter how many sarcomeres are arranged in series, as shown in Fig. 9(a), the net force produced will be equivalent to only one sarcomere. In contrast, if the muscle is organized with the same number of sarcomeres arranged side by side, all can contribute to force production. Thus the correct normalization to make when comparing the isometric strength of different muscles is to relate the force generated to a measure of the number of sarcomeres in parallel; and the muscle dimension that might best reflect this would be the physiological cross-sectional area, i.e. the muscle cross-section at right angles to the fibres that comprise the muscle (group) being studied. Unfortunately, this is a difficult measurement to make in the intact human, since the active muscle fibres acting around any joint have a complex architecture and different angles of pennation. (Moreover, these angles will change as the muscle contracts at different lengths [25,25a].) A reasonable and pragmatic compromise may be to measure the anatomical cross-sectional area of the muscle (group) at right angles to the limb. Before leaving this issue, it should be pointed out that in an isovolume system, such as the muscle fibre, the lateral pressure generated by sarcomeres (in series) during contraction may be transmitted by surrounding structures. In these circumstances the whole muscle may generate more force than might be expected from its cross-sectional area, which reflects the number of sarcomeres in parallel.

Maximal power and muscle size

Muscle power is the product of force and velocity, and velocity is distance covered in unit time. The total distance shortened (hence velocity) by four sarcomeres in series (Fig. 9a) will be four times that of the four sarcomeres in parallel (Fig. 9b). Thus muscle power should be normalized by the number of sarcomeres in series and in parallel. In practice, this might best be reflected by a measurement of muscle volume.

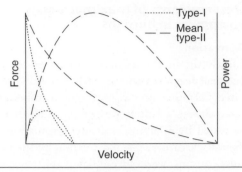

Fig. 10. Force–velocity relationship (continuously declining function) and power–velocity relationship (domed function) for a slow (type-I) and a fast (type-II) population of fibres that generate the same isometric force but whose V_{max} varies in the ratio of 1:4

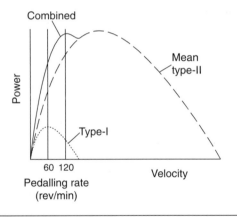

Fig. 11. The component and combined power–velocity relationship for a whole muscle
Based on Fig. 10 and modelled as if composed of two discrete populations of fibres contributing equally to the isometric force (see text). The superimposed pedalling rates are derived from [26a].

Muscle function, body dimensions and performance

The previous section dealt with the physiological principles that need to be taken into account when normalizing maximum muscle function in relation to the size of the active muscle mass. In terms of physical performance, however, it is often useful to be able to express the muscle function in terms of body dimensions when these are the primary determinant of the magnitude of the forces to be overcome. For example, when an old person climbs the stairs to bed, or rises from a low chair, leg muscles operating through lever arms have to raise the body mass; whereas for a sprint cyclist (at least on the flat) the major force to be overcome by the leg muscles is air resistance to the frontal area of the body. Apart from these performance-related expressions, there seems little purpose in seek-

ing to relate muscle function to some 'allometric gold standard' based on body dimension, such as fractional functions of height and weight. Depending upon the pattern of use, the size and intrinsic function of any muscle group can be varied enormously over a wide normal range, quite independently of any body dimensions (and incidentally the function of other muscle groups).

Muscle fibre type and human power output

As indicated earlier in this chapter, the maximal velocity of shortening — and, consequently, the force–velocity relationship — of human muscle fibres is strongly determined by the myosin heavy-chain isoforms expressed. To understand the significance of these effects on human power, it may be useful to refer to a simplified illustration in which the force–velocity relationships of type-I and the *mean* of type-II

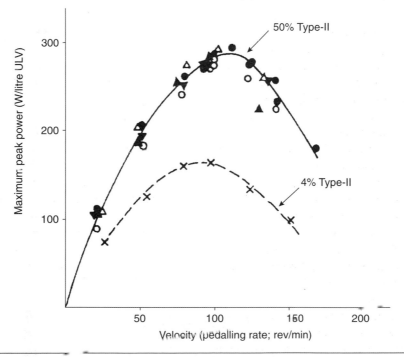

Fig. 12. Experimental data for human maximal peak power during isokinetic cycling in relation to pedalling rate
Power is standardized for upper leg muscle volume (ULV). Data for five subjects (▲, ▼, ○, ●, △) with ~50% type-II fibres (—) and individual data for an ultramarathon runner with only 4% type-II fibres (—×—). (Data taken from [26a]; reproduced with permission from [27].) ©1994, Threme Medical Publishers.

human muscle fibres are compared (Fig. 10). It is important to remember that this is only an illustrative model and a number of simplifications and assumptions are made, albeit based on the literature available (for a full discussion, see [26]). If in a human muscle 50% of the total isometric force were contributed by type-I fibres (regardless of the precise specific strength of different fibre types) then the power–velocity relationship for the whole muscle might be derived by adding those of the two fibre-type populations (Fig. 11). The challenge is to associate these relationships with contraction velocities occurring in human exercise. It has been argued on the basis of experimental data derived from an isokinetic cycle ergometer that the maximum power of the 'combined' power–velocity relationship would approximate to a pedalling rate of ~120 rev/min [26]. It follows from this that the optimal velocity for maximum power of the type-I fibres is around 60 rev/min — beyond that pedalling rate the contribution of type-I fibres to the power output of the whole muscle will diminish. Nevertheless, the V_{max} of these fibres would not be exceeded until pedalling rates in excess of 160 rev/min were attained. An important point to note is the much greater maximum power generated by type-II (mean value) compared with type-I fibres. This is reflected in experimental data when the power–velocity relationship of a group of normal healthy subjects with 50% type-II fibres is compared with that of a highly trained ultra-marathon runner with only 4% type-II fibres in his vastus lateralis (Fig. 12) [27]. The runner's maximum power is only about half of that for the group, even though the data are normalized for upper leg muscle mass.

Human maximum isometric force and fibre type

There has been a debate for many years as to whether, in humans, type-II fibres are intrinsically stronger than type-I fibres. Although the maximum power of human fibres may vary by a factor of ~1:10 between type-I and -IIX fibres, there is much less, if any, difference in the isometric strength when normalized for

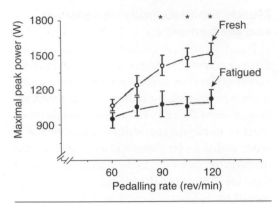

Fig. 13. Human maximal peak power cycling at five different pedalling rates in the fresh (○) and fatigued (●) states
Data given as means ± SEM for six subjects. Statistically significant differences given by *. Adapted with permission from [31].

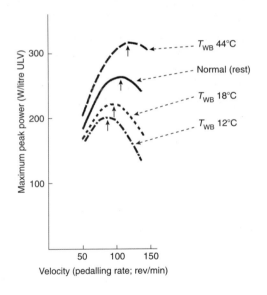

Fig. 14. The relationship of maximal peak power during cycling to pedalling rate under different muscle-temperature conditions
Data are given for normal resting conditions at room temperature and after 45 min immersion in water baths at 44°C, 18°C and 12°C. Arrows indicate the optimal velocity for maximum power which increases with temperature. Abbreviations used: ULV, upper leg muscle volume; T_{WB}, temperature of water bath. Adapted from [31].

cross-sectional area. Indeed, some authors have failed to demonstrate any significant difference, although this could easily be attributed to the difficulties already alluded to of making accurate measurements of muscle-physiological cross-sectional area or fibre-type distribution. If, as seems likely on the basis of the most recent studies, there is a difference it is not as great as that in maximum power but may still be of the order of 1:1.2:1.5, when comparing type-I, -IIA and -IIX human fibres, respectively [13].

> ▶ To generate maximal force a muscle must be fully activated. Most subjects seem able to achieve maximal or near-maximal contractions of the major limb muscles voluntarily.
> ▶ The force generated by a muscle (isometric strength) is dependent on the number of sarcomeres in parallel (cross-sectional area).
> ▶ The maximal velocity of shortening (which determines the power–velocity relationship) of human muscle fibres is determined by the myosin heavy-chain isoforms expressed.
> ▶ Human type-II fibres are intrinsically more powerful than type-I fibres.

Acute plasticity

As pointed out earlier, human muscle exhibits considerable plasticity — the contractile and metabolic properties being modified by long-term changes in, for example, the patterns of activation. It is important to remember, however, that the muscle properties are not fixed even in the short term. They can be substantially modified, not least by the effects of exercise itself. In some circumstances exercise may lead to a potentiation of mechanical output (see, for example, [28]). If the same exercise is prolonged the muscle may fatigue, and the muscle contractile properties may be acutely transformed towards slower characteristics, i.e. the maximal velocity of shortening is reduced and the muscle becomes much less powerful (see Chapter 1 in this volume; see also [29]). In terms of human performance, this means that

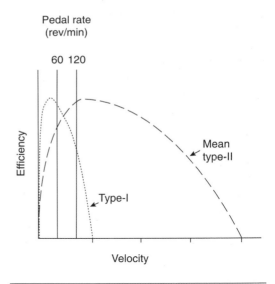

Fig. 15. Schematic illustration of the possible form of the relationship between mechanical efficiency and velocity of contraction for type-I and the mean for type-II muscle fibres
In the absence of systematic data, no relative differences in the maximal efficiencies are given — each type is normalized to the same maximum. The velocity range equivalent to pedalling rates in cycling exercise of 60 and 120 rev/min is derived from Fig. 11.

muscle fatigue may have significantly greater effects at fast movement frequencies than at slow. Thus a standard fatiguing exercise resulted in a ~25% reduction in maximum peak power when cycling at 120 rev/min but had no significant effect at 60 rev/min (Fig. 13) [22].

Changes in muscle temperature brought about either by the metabolic heat of exercise or environmental conditions will also modify contractile and metabolic properties. In humans, as would be expected from isolated muscle experiments, the effect of muscle temperature is velocity dependent (Fig. 14). Thus increases in muscle temperature will make the muscle faster and more powerful at higher velocities. At the same time it may, due to increases in the cross-bridge cycling rate and energy turnover, reduce the economy or efficiency of contractions, and hence the fatigue resistance [30,31]. Clearly these observations have important implications for human sporting performance. Less-often recognized may be the significance of limb and muscle cooling due to impaired circulation for locomotory function in the elderly and some patient populations.

> ▶ The contractile and metabolic properties of human muscle can be altered by exercise and temperature.
> ▶ Increases in muscle temperature will increase the muscle power (and this is velocity dependent) and will increase energy turnover.
> ▶ Reduction in temperature (muscle cooling) has a little-recognized significance for the elderly and patients with locomotory disorders.

Muscle performance and efficiency

The efficiency with which chemical energy is transformed into mechanical energy may be of little significance in human performance requiring a single contraction. However, if exercise is to be sustained for more than a few seconds, mechanical efficiency may be an important determinant of performance, affecting the rate at which substrate is utilized. Unfortunately, relatively little is known about the mechanical efficiency/velocity characteristics of human muscle-fibre types or how these might relate to human exercise. On the basis of animal experiments, it is usually proposed that maximal efficiency occurs at a velocity close to, but slightly less than, that for maximum power (see, for example, [3,32,33]). It has been speculated on the basis of both laboratory and field observations that the efficiency/velocity relationship for the type-I muscle fibres and the average for the type-II muscle fibres may be related to human movement frequencies, as shown in Fig. 15. As has been pointed out, if this speculation is correct, the reciprocal change in efficiencies for the major fibre populations combined with the hierarchy of recruitment could explain the relative constancy of calculated efficiency values for human cycling exercise over a wide range of pedalling rates from ~60 to 110 rev/min [26,33a]. It should be pointed out, however, that the efficiency/velocity relation-

ship is not a fixed property but labile, and will be influenced by muscle temperature as well as factors related to exercise intensity [34].

Summary

A proper understanding of human physical performance depends upon an integration of knowledge derived from many different fields and different types of experiment. Our knowledge at the cellular and molecular levels is increasing rapidly and will doubtless contribute greatly to our understanding of the constraints and limitations to human neuromuscular function, especially in relation to the underlying mechanisms that control the muscle phenotype. Nevertheless, this knowledge only has meaning if it can be integrated with information on how muscle functions in the intact human body to meet the extrinsic demands and conditions imposed upon it. This chapter has sought to highlight the necessity for an integrative approach in which molecular biology, physiology, biochemistry, biomechanics and neurophysiology need to be combined to understand the phenomenon of human movement, whether in the context of top-level sporting performance or, for example, in maintaining or improving the mobility and independence of the elderly.

Conclusions

> ▶ Human exercise performance depends on the generation of muscle power in accordance with the force–velocity and length–tension relationships of the active muscles.
> ▶ Intrinsic muscle power is determined by the number of force-generating units in parallel and in series (as reflected by muscle volume) and the isoforms of the contractile proteins expressed.
> ▶ The myosin heavy-chain isoform expressed is the primary determinant of the maximal velocity of shortening, and hence the power of a muscle fibre. In mature human muscle, there are three main types [type-I (slow); type-IIA (fast); and type-IIX

(fastest)] with many fibres co-expressing variable amounts of the IIA and IIX isoforms.

► The generation of mechanical output is achieved by increasing both the number of motor units recruited (usually with progressive recruitment from smaller to larger units) and the frequency of activation (rate-coding).

► Muscle contractile and metabolic properties may be changed: (a) chronically as a consequence of, e.g., training, aging and immobilization; (b) acutely as a result of, e.g., exercise-induced potentiation and fatigue, and changes in muscle temperature.

► Strategies for maintaining mechanical output and delaying the onset of fatigue will depend on the contractile and metabolic properties of the motor units and the pattern of their recruitment, especially with respect to the rate of energy turnover, i.e. their efficiency, at the contraction velocities required, or chosen, for the task.

References

1. Enoka, R. (1988) *Neuromechanical Basis of Kinesiology*, Human Kinetics Publishers, Champaign

2. Greger, R. and Windhorst, U. (eds.) (1996) *Comprehensive Human Physiology: From Cellular Mechanisms to Integration*, Springer-Verlag, Berlin

3. Goldspink, G. (1992) Cellular and molecular aspects of adaptation in skeletal muscle. In *Strength and Power in Sport* (Komi, P.V., ed.), Human Kinetics Publishers, Champaign

4. Jones, D.A. and Round, J.M. (1991) *Skeletal Muscle in Health and Disease: A Textbook of Muscle Physiology*, Manchester University Press, Manchester

5. Woledge, R.C., Curtin, N.A. and Homsher, E. (1985) *Energetic Aspects of Muscle Contraction*, Academic Press, London

6. Alexander, R.McN. (1984) Elastic energy stores in running vertebrates. *Am. Zool.* 24, 85–94

7. Saltin, B. and Rose, R.J. (eds.) (1994) The racing camel (*Camelus dromedarius*): physiology, metabolic functions and adaptations. *Acta Physiol. Scand.* 150(suppl), 617

8. Ennion, S., Sant'Ana Pereira, J.A.A., Sargeant, A.J., Young, A. and Goldspink, G. (1995) Characterisation of human skeletal muscle fibres according to the myosin heavy chains they express. *J. Musc. Res. Cell Motility* 16, 35–43

9. Larsson, L. and Moss, R.L. (1993) Maximum velocity of shortening in relation to myosin isoform composition in single fibres from human skeletal muscles. *J. Physiol.* 472, 595–614

10. Sant'Ana Pereira, J.A.A., Wessels, A., Nijtmans, L., Moorman, A.F.M. and Sargeant, A.J. (1995) New method for the accurate characterisation of single human skeletal muscle fibres demonstrates a relation between mATPase amd MyHC expression in pure and hybrid fibres types. *J. Musc. Res. Cell Motility* 16, 21–34

11. Sant'Ana Pereira, J.A.A., Sargeant, A.J., de Haan, A., Rademaker, A.C.H.J. and van Mechelen, W. (1996) Myosin heavy chain isoform expression and high energy phosphate content of human muscle fibres at rest and post-exercise. *J. Physiol.* 492(2), 583–588

12. Lowey, S., Waller, G.S. and Trybus, K.N. (1993) Skeletal muscle myosin light chains are essential for physiological speed of shortening. *Nature (London)* 365, 454–456

13. Stienen, G.J.M., Kiers, J.L., Bottinelli, R. and Reggiani, C. (1996) Myofibrillar ATPase activity in skinned human skeletal muscle fibres: fibre type and temperature dependence. *J. Physiol.* 493(2), 299–308

14. Buller, A.J., Eccles, J.C. and Eccles, R.M. (1960) Interactions between motoneurones and muscles in respect of the characteristic speeds of their responses. *J. Physiol.* 150, 417–439

15. Salmons, S. and Vrbová, G. (1969) The influence of activity on some contractile characteristics of mammalian fast and slow muscles. *J. Physiol.* 201, 535–549

16. Unguez, G.A., Bodine-Fowler, S.C., Roy, R.R., Pierotti, D.J. and Edgerton, V.R. (1993) Evidence of incomplete neural control of motor unit properties in cat tibialis after self-reinnervation. *J. Physiol.* 472, 103–125

17. Jones, D.A., Rutherford, O.M. and Parker, D.F. (1989) Physiological changes in skeletal muscle as a result of strength training. *Q. J. Exp. Physiol.* 74, 233–256

18. Henneman, E. and Mendell, L.M. (1981) Functional organization of motoneuron pool and its inputs. In *Handbook of Physiology I*, vol. II, part 1 (Brooks, V.B., ed.), pp. 423–507, American Physiological Society, Bethesda

19. Kernell, D. (1992) Organized variability in the neuromuscular system: a survey of task-related adaptations. *Arch. Ital. Biol.* 130, 19–66

20. Greig, C.A., Hortobagyi, T. and Sargeant, A.J. (1985) Quadriceps surface e.m.g. and fatigue during maximal dynamic exercise in man. *J. Physiol.* 369, 180P

21. Ivy, J.L., Chi, M.-Y., Hintz, C.S., Sherman, W.M., Hellendall, R.P. and Lowry, O.H. (1987) Progressive metabolic changes in individual human muscle fibers with increasing work rates. *Am. J. Physiol.* 252, C630–C639

22. Beelen, A. and Sargeant, A.J. (1991) Effect of fatigue on maximal power output at different contraction velocities in humans. *J. Appl. Physiol.* 71(6), 2332–2337

23. Beelen, A., Sargeant, A.J., Lind, A., de Haan, A., Kernell, D. and van Mechelen, W. (1993) Effect of contraction velocity on the pattern of glycogen depletion in human muscle fibre types. In *Neuromuscular Fatigue* (Sargeant, A.J. and Kernell, D., eds.), pp. 93–95, North Holland, Amsterdam

23a. Greig, C.A., Sargeant, A.J. and Vollestad, N.K. (1985) Muscle force and fibre recruitment during dynamic exercise in man. *J. Physiol.* 371, 176P

23b. Beelen, A., Sargeant, A.J., Jones, D.A. and de Ruiter, C.J. (1995) Fatigue and recovery of voluntary and electrically elicited dynamic force in humans. *J. Physiol.* 484, 227–235

24. James C., Sacco, P. and Jones, D.A. (1995) Loss of power during fatigue of human leg muscles. *J. Physiol.* 484, 237–246

24a. Maganaris, C.N., Baltzopoulos, V. and Sargeant, A.J. (1998) Differences in human antagonistic ankle dorsiflexor coactivation between legs; can they explain the moment deficit in the weaker plantar flexor leg? *Exp. Physiol.* 83, 843–855

25. Narici, N.V., Binzoni, T., Hiltbrand, E., Fasel, J., Terrier, F. and Cerretelli, P. (1996) *In vivo* human gastrocnemius architecture with changing joint angle at rest and during graded isometric contraction. *J. Physiol.* 496(1), 287–297

25a. Maganaris, C.N., Baltzopoulos, V. and Sargeant, A.J. (1998) In vivo measurments of the triceps surae complex architecture in man: implications for muscle function *J. Physiol.* 512, 603–614

26. Sargeant, A.J. and Jones, D.A. (1995) The significance of motor unit variability in sustaining mechanical output of muscle. *Adv. Exp. Med. Biol.* **384**, 323–338

26a. Sargeant, A.J., Hoinville, E. and Young, A. (1981) Maximum leg force and power output during short-term dynamic exercise. *J. Appl. Physiol.* **51(5)**, 1175–1182

27. Sargeant, A.J. (1994) Human power output and muscle fatigue. *Int. J. Sports Med.* **15(3)**, 116–121

28. de Ruiter, C.J., de Haan, A. and Sargeant, A.J. (1996) Fast twitch muscle unit properties in different muscle compartments. *J. Neurophysiol.* **75**, 2243–2254

29. de Haan, A., Jones, D.A. and Sargeant, A.J. (1989) Changes in power output, velocity of shortening and relaxation rate during fatigue of rat medial gastrocnemius muscle. *Pfluegers Archiv. Eur. J. Physiol.* **413**, 422–428

30. Edwards, R.H.T., Harris, R.C., Hultman, E., Kaijser, L., Koh, D. and Nordesjo, L.-O. (1972) Effect of temperature on muscle energy metabolism and endurance during successive isometric contractions sustained to fatigue of the quadriceps muscle in man. *J. Physiol.* **220**, 335–352

31. Sargeant, A.J. (1987) Effect of muscle temperature on leg extension force and short-term power output in humans. *Eur. J. Appl. Physiol.* **56**, 693–698

32. Lodder, M.A.N., de Haan, A. and Sargeant, A.J. (1991) Effect of shortening velocity on work output and energy cost during repeated contractions of the rat EDL muscle. *Eur. J. Appl. Physiol.* **62**, 430–435

33. Rome, L.C. (1993) The design of the muscular system. In *Neuromuscular Fatigue* (Sargeant, A.J. and Kernell, D., eds.), pp. 129–136, North Holland, Amsterdam

33a. Sargeant, A.J. and Rademaker, A. (1996) Human muscle fibre types and mechanical efficiency during cycling. In *The Physiology and Pathophysiology of Exercise Tolerance* (Steinacker, J.M. and Ward, S.A., eds.), pp. 247–251, Plenum, New York

34. Zoladz, J.A., Rademaker, A. and Sargeant, A.J. (1995) Oxygen uptake does not increase linearly with power output at high intensities of exercise in humans. *J. Physiol.* **488(1)**, 211–218

3

Muscle energetics: aerobic strategies

D.L. Turner*[1] and H. Hoppeler†

*Sport and Exercise Research Centre, School of Applied Sciences, South Bank University, 103 Borough Road, London SE1 0AA, U.K., and †Department of Anatomy, University of Bern, Bühlstrasse 26, CH-3000 Bern 9, Switzerland

Introduction

Resting or basal energy expenditure is necessary to maintain various bodily functions, including ionic gradients, turnover of basic cell constituents (such as protein, carbohydrate and fats), thermoregulation and posture. Increased physical activity can evoke great changes in many energy-expending functions and, thereby, necessitates large increases in ATP hydrolysis and whole-body energy expenditure.

The aim of this chapter is to describe the basic mechanisms by which muscles can generate this energy aerobically during dynamic activity, up to and including the limits of maximal oxygen utilization ($\dot{V}_{O_2max}$).

Cellular energy turnover during dynamic exercise

Energy utilization during the contraction–relaxation cycle

Muscular contraction involves the active detachment–attachment–detachment of actin and myosin proteins and is an endergonic reaction. The free energy needed for this cycle to proceed is released by the cleavage of the high-energy bond between the two terminal phosphate groups of ATP and is catalysed by myosin ATPase. This 'exergonic' process is extremely fast (in the order of milliseconds) and can increase energy production by up to 1000 times the resting or basal rate. An enhanced rate of ATP hydrolysis is also essential for the

increased activity of two other ATPases in contracting muscle. The first is involved in the active re-uptake of Ca^{2+} ions from the contractile apparatus into the sarcoplasmic reticulum so that relaxation can occur. The second is involved in maintaining trans-sarcolemmal ionic gradients and, therefore, the cell membrane potential. Contractile processes may account for approximately 70% of the ATP hydrolysis measured during muscular contraction, the remainder being necessary for Ca^{2+} pumping and maintenance of ionic gradient [1].

Replenishment of energy utilized during contraction and relaxation

The role of high-energy phosphates
The total intracellular concentration of ATP in muscle is low (approximately 8 mM) and would only sustain a dozen or so muscle contractions were it not replenished. Therefore, ATP must be rapidly resynthesized 'on site' from its breakdown products, ADP and inorganic phosphate (P_i), at a rate commensurate with its degradation, for repeated muscle-contraction cycles to continue. During muscular contraction and relaxation the degradation of ATP at the actomyosin cross-bridge is intimately coupled to the degradation of another high-energy phosphate, phosphocreatine (PCr), via the reaction catalysed by an isozyme of creatine kinase (M-line linked creatine kinase, M-CK). The coupled reactions for PCr and ATP degradation are:

$$ATP \rightarrow ADP + P_i + energy \qquad (1)$$
$$PCr + H^+ + ADP \leftrightarrow Creatine + ATP \qquad (2)$$

$$PCr + H^+ \rightarrow Creatine + P_i + energy \qquad (3)$$

---

[1] To whom correspondence should be addressed.

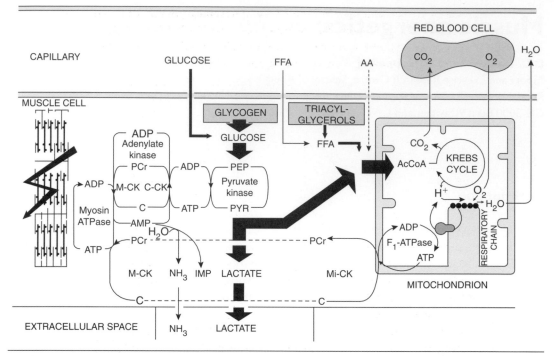

Fig. 1. Schematic drawing of pathways for energy and substrate supply in muscle cells

Abbreviations used: Mi-CK, mitochondrial form of creatine kinase; M-CK, M-line bound form of creatine kinase; C-CK, cytosolic creatine kinase; AcCoA, acetyl-coenzyme A; PCr, phosphocreatine; C, creatine; FFA, free fatty acids; PYR, pyruvate; PEP, phosphoenolpyruvate; AA, amino acids. The widths of the lines are approximations of the maximal flux rates. The zig-zag line across the *muscle cell represents the release of energy, through active contraction of muscle fibres, that can be used to move body mass. During exercise, the flux rates may change in order to accommodate the increase in myosin ATPase activity. Reproduced from [2] with permission. ©1991 Company of Biologists.*

The degradation of PCr, and its subsequent resynthesis, are in fact integral components of aerobic ATP turnover or, in more general terms, the chemical energy flow from substrate stores to the actomyosin cross-bridge (Fig. 1; [2]). Actually, the intracellular concentration of ATP only decreases by approximately 20% during aerobic exercise, even at $\dot{V}_{O_2max}$, because it is buffered so well by the 'on site' breakdown of PCr (Table 1). On the other hand, the microenvironment surrounding M-CK may have large activity-induced changes in energy charge (EC):

$$EC = ([ATP] \cdot 0.5[ADP]) \cdot ([ATP] + [ADP] + [AMP])^{-1} \qquad (4)$$

These changes presumably drive the reversible reaction (eqn 2) to the right, via very rapid changes in [ATP], [ADP] and/or [AMP]. The maintenance of an optimal, high cellular EC ratio may be important, because the free energy liberated from ATP hydrolysis is thought to be directly proportional to EC [3].

An alternative way to maintain a high EC is to degrade ADP to AMP (a reaction catalysed by adenylate kinase), and thence to non-regulatory compounds such as inosine monophosphate (IMP), ammonia (NH_3^+) or hypoxanthine via the reactions catalysed initially by AMP deaminase (Fig. 1 and Table 1). During heavy aerobic exercise (>80% $\dot{V}_{O_2max}$) or prolonged exercise at lower intensities, a fall in EC and the subsequent increases in IMP, NH_3^+ and hypoxanthine may indicate the existence of a general imbalance between the rates of ATP resynthesis and utilization (Fig. 2). Alternatively, the build up of certain metabolites

Table 1. The changes in high-energy phosphate metabolites, lactate, ammonia, glycolytic intermediates and NADH [mmol·(kg of dry mass)$^{-1}$] within human vastus lateralis muscle biopsies taken during 3.5–10 min of exercise at each intensity [3,6]

	Rest	40–50% O_2max	60–80% O_2max	100% O_2max
PCr	83 ± 2	$74 \pm 2^*$	$36 \pm 4^*$	$17 \pm 2^*$
ΔP_i	0	9.3 ± 1.1	$42.3 \pm 3.9^*$	$59.7 \pm 4.0^*$
ATP	26 ± 1	25 ± 1	25 ± 1	$20 \pm 1^*$
ADP	3.1 ± 0.1	3.3 ± 0.1	$3.6 \pm 0.1^*$	$3.8 \pm 0.2^*$
AMP	0.1 ± 0.01	0.1 ± 0.01	0.1 ± 0.01	0.2 ± 0.03
IMP	<0.01	<0.01	0.26 ± 0.06	$3.5 \pm 0.51^*$
TAN	29 ± 1	29 ± 1	29 ± 1	$24 \pm 1^*$
TAN + IMP	29 ± 1	29 ± 1	29 ± 1	28 ± 1
EC	0.95	0.92	0.92	0.92
NH_3^+	0.5 ± 0.1	0.5 ± 0.1	–	$4.1 \pm 0.5^*$
Glycogen^a	284	234	222	141
Lactate	1.8 ± 0.2	2.7 ± 0.6	$33.5 \pm 5.1^*$	$100.0 \pm 5.5^*$
Pyruvate	0.13 ± 0.01	0.19 ± 0.02	$0.46 \pm 0.02^*$	$0.46 \pm 0.04^*$
Glucose	1.3 ± 0.3	0.4 ± 0.2	2.3 ± 0.5	$9.9 \pm 0.5^*$
G-6-P	0.8 ± 0.2	$1.7 \pm 0.3^*$	$2.9 \pm 0.3^*$	$6.5 \pm 0.3^*$
NAD^+	1.9 ± 0.1	–	–	2.0 ± 0.1
NADH	0.2 ± 0.02	$0.1 \pm 0.01^*$	$0.3 \pm 0.03^*$	$0.3 \pm 0.04^*$

Values are given as mean $\pm$ SEM, n = 8–10. *Indicates significant difference from resting values P<0.05. Abbreviations used: IMP, inosine monophosphate; TAN, total adenine nucleotides; G-6-P, glucose 6-phosphate. ^{a}Indicates mM glucosyl units·(kg of dry mass)$^{-1}$.

(ADP, AMP, P_i and creatine) may be necessary for modulating the roles of various cellular metabolic pathways to preserve cellular respiration and thus the minimal breakdown of cellular ATP [3].

The role of oxidative metabolism

Normal, everyday activity rarely utilizes just one type of metabolic replenishment of ATP. The balance between using (i) solely the breakdown of high-energy phosphates such as ATP itself or PCr, (ii) anaerobic and (iii) aerobic production of ATP, depends on several variables including intensity, duration and type of exercise (see Table 1 and Fig. 2). Aerobic metabolism can be viewed as the predominant, but not the sole, source of ATP generation during pro-

longed (>2 min), mild-to-moderate-intensity exercise with adequate muscle perfusion.

Oxidative rephosphorylation of ADP within subcellular organelles called mitochondria allows resynthesis of ATP, which in turn replenishes the intracellular pool of PCr. This will maintain the operation of the PCr circuit, or 'energy shuttle', between mitochondria and contractile filaments (Fig. 1). The initial replenishing cycle in the mitochondria is catalysed by another compartmentalized CK isozyme (mitochondrial creatine kinase; Mi-CK). Mi-CK catalyses energy transfer across the mitochondrial membranes into the sarcoplasm, where yet further CK isozymes within the cytosol (Fig. 1) catalyse cyclical reactions. In this way, these serially arranged cyclical meta-

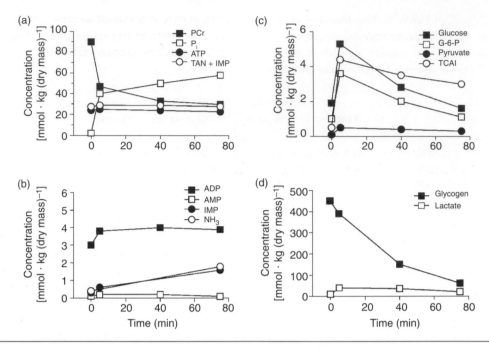

Fig. 2. Changes in concentration of muscle high-energy phosphates, glycolytic intermediates and tricarboxylic acid cycle intermediates (TCAI) during prolonged exercise at approximately 60–80% $\dot{V}_{O_2max}$

Changes in concentration of (a) PCr, P_i, ATP and TAN + IMP, (b) ADP, AMP, IMP and NH_3, (c) glucose, G-6-P, pyruvate and TCAI, and (d) glycogen and lactate.

Abbreviations used: TAN = ATP + ADP + AMP; IMP (inosine monophosphate); NH_3, ammonia; G-6-P, glucose 6-phosphate. Data taken from [3,6].

bolic 'cogs' complete the multiple-step transfer of chemical energy from mitochondria to the actomyosin cross-bridge.

During mild exercise, oxidative rephosphorylation of ADP to ATP can occur simultaneously with significant cytosolic glycolysis and lactate accumulation (Table 1). Increased glycolysis may be mediated by increases in intracellular calcium concentration or by increases in intracellular regulatory metabolites (ADP, AMP, NADH, etc.) rather than severe tissue hypoxia *per se* [4]. Both the increase in substrate phosphorylation via anaerobic glycolysis and the breakdown of intracellular PCr could be triggered by increased H^+ and ADP production (eqn 2) during aerobic exercise simply as a result of an increased rate of ATP turnover. This does not necessarily mean, therefore, that there is a dysfunction in cellular respiration, merely that there are shifts in metabolic pathway control, with both glycolysis and

respiratory-chain phosphorylation occurring in a concerted fashion during exercise.

There are some important prerequisites for aerobic metabolism to function efficiently during muscular exercise. First, there must be an adequate flow of oxygen from the atmosphere to the exercising muscle mass through a large series of resistive steps of the oxygen-transport pathway, in order to maintain the minimal concentration of oxygen necessary for mitochondrial respiration [5]. The partial pressure of oxygen in the vicinity of the terminal electron-chain units is thought to be near to zero during exercise, thereby maintaining the overall gradient for oxygen diffusion: the mitochondria act as the ultimate sink for oxygen.

Secondly, there must be an adequate supply of substrates, such as carbohydrate, fat, protein, acetyl coenzyme A (acetyl-CoA) and NADH/FADH2, to maintain the appropriate level of aerobic glycolysis, fat oxidation, protein breakdown, tricarboxylic acid (TCA) cycling and electron transfer/oxidative phosphorylation.

Thirdly, there must be sufficient disposal of potentially deleterious metabolic reaction products (carbon dioxide, excess protons and metabolic H_2O) and other by-products of increased metabolism, such as heat. All three of these prerequisites involve intricate and complex interaction of active muscles with other systems of the body.

▶ Muscular contraction is an endergonic reaction that obtains its free energy from the hydrolysis of ATP, a reaction catalysed by myosin ATPase.
▶ To enable repeated muscle contraction cycles to continue, ATP is resynthesized close to the site of utilization.
▶ Degradation of ATP is coupled to the degradation of PCr during muscular contraction.
▶ During heavy aerobic exercise, there may be an imbalance between the rates of ATP utilization and resynthesis.
▶ Oxidative rephosphorylation of ADP to ATP in the mitochondria occurs during prolonged mild-to-moderate-intensity exercise.
▶ For aerobic metabolism to function efficiently during muscular exercise, there must be: (i) adequate oxygen flow to the exercising muscle; (ii) adequate supply of substrates for glycolysis, fat oxidation, protein breakdown, etc.; (iii) sufficient disposal of deleterious reaction by-products.

Biochemical basis of aerobic ATP replenishment in skeletal muscle and its topological arrangement

Biochemical pathways

The breakdown of substrates either from intramuscular stores or from the blood ultimately yields large quantities of ATP via aerobic metabolism. Oxygen is the terminal acceptor of protons from cytochrome oxidase in the respiratory-chain complex.

Glycogen and plasma glucose utilization

Glycogen is stored intramuscularly as granules and can be sequentially broken down to form pyruvate via the Embden–Meyerhof pathway by enzymes which may be in discrete topological arrangements within the cytoplasm [6]. Blood-borne glucose can be fed into this pathway, once it has been 'trapped' intracellularly by phosphorylation to glucose 6-phosphate, which is catalysed by the enzyme hexokinase. The energy balance sheet for complete aerobic breakdown of glycogen or glucose is given in the box below.

The majority of ATP produced by aerobic glycolysis is via oxidation of NADH by oxygen within the respiratory chain. The rate of glycogen breakdown is increased with increasing intensity of exercise (Table 1), but at any given submaximal workload its degradation is greatest in the initial stages and declines curvilinearly with time (Fig. 2) [7]. There is a high correlation between the time to exhaustion dur-

---

Glycolytic yield
= −1ATP	(glucose + 1ATP → glucose 6-phosphate + ADP)
−1ATP	(fructose 6-phosphate + ATP → fructose 1,6-bisphosphate + ADP)
+4ATP	(direct cytoplasmic generation)

Aerobic yield
= +6ATP	(mitochondrial oxidation of cytoplasmic NADH)
+24ATP	(mitochondrial oxidation of mitochondrial NADH)
+4ATP	(mitochondrial oxidation of mitochondrial FADH)
+2ATP	(mitochondrial generation via GTP formation)

Total
= +38ATP	(glucose)
or +39ATP	(glycogen)

ing submaximal exercise at between 60% and 80% $\dot{V}_{O_2max}$ and the near-complete depletion of intramuscular glycogen stores in humans measured biochemically [8] or morphologically after ultra-endurance exercise [9]. This does not mean that there is a true cause–effect relationship between the depletion of intramuscular glycogen stores and performance, but does signify the importance of intramuscular glycogen to aerobic exercise.

As intramuscular glycogen utilization diminishes during submaximal exercise (60–80% $\dot{V}_{O_2max}$), adequate ATP replenishment is maintained by increased muscle uptake of blood glucose, with concomitant increases in liver glycogenolysis and facilitative glucose diffusion across the muscle sarcolemma via a glucose transporter (GLUT 4; [10]). There is also evidence to suggest that muscles may take up lactate from the circulation and oxidize it via its conversion to pyruvate. There may also be a minor utilization of protein breakdown to fuel oxidative metabolism and to replenish or elevate tricarboxylic acid cycle intermediates (TCAI; Fig. 2).

Triacylglycerol and free fatty acid utilization
Prolonged exercise at <80% $\dot{V}_{O_2max}$ is characterized by minimal alterations in ATP, a low intensity of exercise, reduced glycogen utilization, increased utilization of blood glucose and increased fat oxidation. For example in ultra-marathon runners, there is a gradual decrease in respiratory exchange ratio [11] indicative of a greater fat utilization. The seemingly obligatory decrease in the percentage of $\dot{V}_{O_2max}$ at which subjects can perform very prolonged exercise may be causally linked to the processes of fat mobilization, transport and utilization, limiting the maximal rate of mitochondrial rephosphorylation of ADP. The energy available from fat stored in muscle or adipose tissue is greater than glycogen, so the duration of exercise may be much greater during periods of fat oxidation, albeit at a lower intensity.

The intramuscular store of oxidizable fat is in the form of triacylglycerol (TAG) droplets. These are often found in close proximity to mitochondria and are, therefore, related to the metabolic characteristics of the fibre type. This

store is relatively small, typically approx. 2% of the muscle fibre volume [12] and can not be solely responsible for the prolonged duration of some physical activities. This is achieved by increased TAG mobilization from adipose tissue, near to the active muscle mass, often between muscles and also at more distant locations. The result of TAG breakdown in adipocytes is an increase in the circulating pool of TAG, which is transported as free fatty acids (FFAs) in association with albumin or as TAG in chylomicrons to the active muscle cells [2]. Lipoprotein lipase (LPL) activity has been shown to be increased by pre- and post-transcriptional mechanisms in muscle cells which are actively contracting. Although the mechanism is unknown, it may involve a cyclic AMP/ protein kinase cascade. Activated LPL is then exported to the luminal surface of capillary endothelial cells where it hydrolyses circulating TAG in chylomicrons and very-low-density lipoproteins. The FFAs are then transported through the interstitial space, probably by albumin or an active transporter. Transport across the sarcolemma is thought to be mediated via an intricate physico-chemical equilibrium between the extracellular FFA–albumin complex, the intracellular FFA-binding proteins (FABP) and the membrane lipid phase. Liberation of FFA from intramuscular TAG droplets is accelerated during exercise by hormone-sensitive lipase, which is thought to be activated in tandem with LPL [2]. Intracellular transport of FFAs, from both the sarcolemma and from TAG droplets, to the mitochondria is probably by a series of FABPs which increase the effective solubility of the normally hydrophobic FFAs [2,13]. The balance between intra- and extra-cellular sources of FFAs may change with time, fibre type, exercise intensity, physical conditioning, etc. The mechanism(s) for this phenomenon are unknown.

Mitochondrial respiration
Mitochondria are the organelles that act as a common collection point for substrates necessary for TCA cycling and electron transfer/ oxidative phosphorylation. Structurally, mitochondria are composed of an outer bilipid and

inner bilipid membrane, the latter surrounding the inner mitochondrial matrix. Many of the enzymes responsible for fat oxidation, TCA cycling and $NADH/FADH_2$ production are found in complex structural and functional associations in the mitochondrial matrix, often existing in 'metabolons', and appear to exist in relatively strict concentration ratios [14].

Substrate entry: Pyruvate from glycogen (glucose) breakdown is transported into mitochondria by specific translocases and then decarboxylated to acetyl-CoA by the pyruvate dehydrogenase system. This system is a multienzyme complex which is located within the inner mitochondrial membrane and is activated by pyruvate and Ca^{2+}, but deactivated by high ATP/ADP, $NADH/NAD^+$ and acetyl-CoA/CoA ratios.

FFAs are attached to CoA to form acyl coenzyme A (acyl-CoA) by the enzyme acyl synthase on the outer mitochondrial membrane. The acyl-CoA is then complexed with carnitine and transported into the mitochondrial matrix by the carnitine acyl transferase system. The FFA carbon skeleton is then progressively cleaved by β-oxidation catalysed by a number of enzymes in the mitochondrial matrix to produce copious acetyl-CoA units.

The control mechanism by which the relative proportions of acetyl-CoA units derived from either glycolysis or fat oxidation are regulated during exercise is unknown. One theory states that oxidation of FFAs inhibits the oxidation of carbohydrate by increasing the concentration of cytosolic citrate due to leakage from the mitochondria. Citrate is a potent inhibitor *in vitro* of phosphofructokinase (PFK), an enzyme within the Embden–Meyerhof pathway. A consequence of lower PFK activity is an accumulation of glucose 6-phosphate, which in turn decreases glucose utilization by inhibiting hexokinase. This potential mechanism, the glucose–fatty acid cycle, has been criticized recently, because increases in cytosolic citrate and glucose 6-phosphate were not found during prolonged exercise, despite decreased glycogen breakdown, reduced glucose utiliza-

tion and increased FFA oxidation, following endurance training [15]. Alternatively, fat oxidation may increase the acetyl-CoA/CoA ratio, thereby reducing the activity of the pyruvate dehydrogenase system and thus aerobic glycolysis.

TCA cycle (Krebs cycle): Whether the acetyl-CoA units come from aerobic glycolysis or from β-oxidation of FFAs, they are used to drive the TCA reactions within mitochondria. Oxaloacetate is condensed with the acetyl-CoA to form citrate: this reaction is catalysed by citrate synthase, an enzyme whose activity is often taken as an index of overall aerobic metabolism. After transformation of citrate to isocitrate, isocitrate dehydrogenase cleaves one carbon (released as carbon dioxide) and two hydrogen atoms (picked up by NAD^+ to form NADH) resulting in the formation of α-ketoglutarate. Further carbon and hydrogen atoms are cleaved in reactions which produce some ATP via GTP formation, leaving succinate, which in turn is oxidized to fumarate by succinate dehydrogenase. The hydrogen atoms from the latter step are bound to FAD^+ to form $FADH_2$, which is an integral part of the respiratory chain (complex II, see later) residing on the inner mitochondrial membrane. Water is added to fumarate and the resulting malate is further dehydrogenized to oxaloacetate, closing the cycle. Amino acids may be fed into various points of the TCA cycle via anaplerotic reactions involving either the purine nucleotide system (reamination of IMP) or pyruvate, some at strategic points such as α-ketoglutarate, fumarate, oxaloacetate and malate or as acetyl-CoA (from pyruvate). The importance of increasing TCAIs during prolonged exercise (Fig. 2) is not known for certain, but it has been suggested that increased substrate concentrations will increase the flux through the TCA cycle as a whole [16,17]. Some aerobic glycolysis and amino acid breakdown, therefore, seem necessary during prolonged exercise, even though FFA utilization is the predominant mechanism supporting aerobic ADP rephosphorylation in the mitochondria.

Electron/proton transfer and ADP rephosphorylation: The inner mitochondrial membrane is highly impermeable to protons and so a powerful proton electrochemical gradient is set up by extrusion of the protons, carried by NADH and $FADH_2$ into the intermembrane space. This extrusion process is performed by four respiratory-chain complexes which straddle the inner mitochondrial, bilipid membrane. The three biggest complexes, namely complex I (NADH:ubiquinone oxidoreductase), complex III (ubiquinol:cytochrome-*c* oxidoreductase) and complex IV (cytochrome-*a* reductase and cytochrome-a_3 oxidase) are aggregations of multiple protein subunits (13–25 units). Complex II is the FAD^+:succinate dehydrogenase complex which is also part of the TCA cycle. Protons are accepted from NADH on the inner surface of the membrane (matrix side) and either carried by cofactors (e.g. ubiquinone) or pumped (e.g. subunit III of cytochrome-*c* oxidase) to the outer surface of the inner-membrane bilipid layer and released, with the concomitant movement of electrons through the respiratory chain units via suitable carriers (e.g. cytochrome oxidase or iron–sulphur proteins). The last link in the respiratory chain involves the production of metabolic water from oxygen moieties (0.5 O_2) and protons catalysed by cytochrome-a_3 oxidase. The net reaction results in oxidation of NADH and $FADH_2$ to NAD^+ and FAD^+ by oxygen. NAD^+ and FAD^+ must be subsequently reduced again, by either glycolysis or TCA cycling to continue respiratory chain function.

The electrochemical gradient that is produced by the proton extrusion is rapid enough and has sufficient free energy to drive ATP production via the F_1F_0-type ATP synthase enzyme complex, which is also situated in the inner mitochondrial membrane [2,18]. Mitochondria thus serve the vital function of linking aerobic glycolysis or fat oxidation, which produce NADH and $FADH_2$, to oxidative rephosphorylation of ADP by the respiratory chain units. The respiratory chain units and the F_1F_0-ATPase, which acts as a channel for the protons to move back into the mitochondrial matrix, are spatially juxtaposed. These two major structural elements are present in well-defined molar ratios and are densely packed into the inner membrane of fully developed mitochondria. The capacity of a mitochondrion for oxidative phosphorylation must, therefore, be limited by the surface area of the inner mitochondrial membrane, assuming the packing density of the elements is already maximal.

The energy contained in the intramitochondrial ATP produced by oxidative phosphorylation is used to resynthesize PCr (eqn 2), which is enzymically catalysed by Mi-CK. In this way, the cellular pool of PCr can be replenished by oxidative metabolism within mitochondria.

▶ The maximal contractile rate of skeletal muscle is set by the reaction rate of myosin ATPase. The supply of ATP for this reaction is adequately met by ATP production through the M-CK reaction, which may in turn be met by anaerobic sources of ATP or aerobic metabolism.

▶ The maximal rate of aerobic exercise is set by the rate at which oxidative metabolism can be maintained in the mitochondria (and therefore maintenance of the PCr shuttle). This rate is lower than both the maximal contractile rate (M-CK activity) and maximal rate of anaerobic PCr degradation and so it is the respiratory machinery (from lung to mitochondrial respiratory chain unit) and not the contractile machinery that set the limit for aerobic capacity, i.e. $\dot{V}_{O_2max}$.

Morphological correlates of biochemical pathways

There is now a great deal of information on the morphological correlates of biochemical pathways supporting oxidative metabolism in muscle. Techniques involving light and electron microscopy, stereology and histochemical and immunocytochemical staining have allowed a detailed map of the muscle cell to be constructed (Fig. 1). The findings have helped significantly in allowing the biochemical pathways to be interpreted with respect to cellular energy turnover and its control during exercise.

In simple terms, a unit volume of mitochondrial protein may be viewed as having a relatively invariable maximal capability to produce ATP. This is because in skeletal muscles there is a relatively constant ratio between inner mitochondrial membrane surface area and mitochondrial volume: approximately 20–40 $m^2 \cdot ml^{-1}$ [19]. This constancy is useful, because it means that either measuring biochemically the enzyme activity of an important oxidative enzyme, such as citrate synthase, or morphometrically measuring the volume of mitochondria in a unit volume of muscle fibre, will yield an indication of skeletal muscle oxidative ADP rephosphorylation capacity. Of course, submaximal rates — and possibly maximal rates — of mitochondrial ATP production may be affected by a number of factors involved in respiratory control, such as substrate availability of ADP, P_i and NADH, ADP translocase or Mi-CK activity and temperature [20]. The evidence available, both correlative and biochemical, suggests that a unit volume of mitochondria are operating at 60–80% of the maximal rate of respiration ($\dot{V}_{mito,max}$) during maximal exercise ($\dot{V}_{O_2max}$). This being the case, $\dot{V}_{O_2max}$ would be set by the volume of mitochondria, and a change in $\dot{V}_{O_2max}$, for example after endurance training, would be brought about mainly by increasing the number of active mitochondria. Other evidence suggests that mitochondria are not operating at (near) maximal rates but at something like 30–50% $\dot{V}_{mito,max}$: this corresponds to the apparent K_m for respiratory control [21]. In this case, an increase in $\dot{V}_{O_2max}$ could be brought about by increasing the rate of respiration in each individual mitochondrion and/or the number of active mitochondria (respiratory-chain aggregations). Whichever is the case, $\dot{V}_{O_2max}$ is correlated to the volume of mitochondria measured from muscle biopsy material, both within different human populations and between different non-human species, with respect to body size or exercise performance [22].

There are three main fibre types in human muscle: type-I (slow twitch, oxidative), type-IIA (fast twitch, oxidative–glycolytic) and type-IIB (fast twitch, glycolytic). With respect to oxidative enzyme activities, measured as maximal enzyme activity rates and mitochondrial volume density estimates from stereological analyses *in vitro*, the three fibre types have different levels of oxidative capacity: type-IIB < type-IIA < type-I. The vast majority of mitochondria are found in type-I and type-IIA fibre and not in type-IIB fibres and it is predominantly type-I and type-IIA fibres that are used in aerobic exercise, up to and including $\dot{V}_{O_2max}$ in humans [8]. Whether this coincidence is purely fortuitous or a design feature is not known.

There is a heterogeneous distribution of mitochondria within a muscle fibre. Below the sarcolemmal membrane there are clusters of tightly juxtaposed mitochondria in the vicinity of neighbouring capillaries, whereas in subsarcolemmal areas away from capillaries, there is a very sparse concentration of mitochondria. Mitochondria also appear between myofibrils, with a tapering frequency from the subsarcolemmal region in towards the core of the muscle fibre [23]. The absolute volumes vary roughly with the fibre type, such that type-I and some type-IIA fibres have elevated volume densities of mitochondria at all locations compared with type-IIB fibres. Mitochondria viewed in three dimensions show considerable complexity in morphology. The degree of complexity again seems loosely correlated to volume density and fibre type [23]. In extremely oxidative fibres, contiguous mitochondria may form functionally continuous pathways for oxygen from the periphery to the centre of the muscle fibre, leading to the supposition that there may be a structural continuity, i.e. a mitochondrial reticulum. The reason for this seemingly complicated mitochondrial patterning in muscle fibres is at present not fully understood, although it may be involved in optimizing oxygen flow from capillaries into the regions of myocytes with fewer mitochondria [23]. There is recent information that suggests that the different anatomically situated clusters of mitochondria have different biochemical capacities for oxidative ADP rephosphorylation, i.e. different packing density of respiratory chain units (interfibrillar capacity > subsarcolemmal

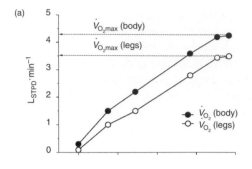

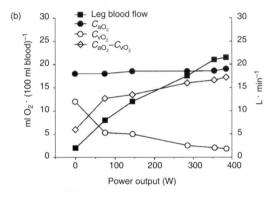

Fig. 3. Measurements of (a) whole-body and leg $\dot{V}_{O_2}$ and calculated $\dot{V}_{O_2max}$ and (b) leg blood flow and oxygen concentrations in arterial and draining venous blood during cycle ergometry in humans

Values for single leg blood flow are doubled to yield total leg blood flow. Abbreviations used: L, litre; L_{STPD}, Litres, standard temperature and pressure, dry. C_{aO_2}, arterial O_2 concentration; C_{vO_2}, muscle venous O_2 concentration. Data taken from [26].

capacity) and they may actually be at different stages of biogenesis [24].

> ▶ The rate of mitochondrial ATP production may be affected by: availability of ADP, P_i and NADH; activity of ADP translocase or Mi-CK; and temperature.
> ▶ The three main types of fibre in human muscle have different levels of oxidative capacity: type-IIB < type-IIA < type-I.

> ▶ The distribution of mitochondria within a muscle fibre varies according to the fibre type.
> ▶ $\dot{V}_{O_2max}$ is correlated with the volume of mitochondria within a unit volume of muscle fibre.

Oxygen and substrate delivery

Functional aspects

The intramuscular supplies of oxygen (in solution or bound to myoglobin) and oxidizable substrates are quantitatively small. In order that exercise may be continued beyond store depletion, oxygen and substrates must be delivered to the active muscle fibres by the cardiovascular (microvascular) system. The central cardiovascular response to prolonged exercise is an increase in cardiac output as a result of increases in heart rate and also stroke volume. There is a general, sympathetically mediated vasoconstriction in the periphery, particularly in the splanchnic and renal regions. In contrast, a potent vasodilation occurs in active skeletal muscle, which is mediated by a number of 'vasoactive agents' released during exercise. The magnitude of the central cardiovascular response is related to the mass of active muscle and the intensity of exercise [25]. In active muscle, it has been shown repeatedly that local blood flow increases substantially — primarily as a function of decreased peripheral resistance and, to a minor extent, to an increase in perfusion pressure [26,27]. This implies that the number of capillaries that are perfused must increase substantially. Capillaries do not respond to vasoactive substances, which must instead exert their effect mostly by reducing smooth muscle tone in small arterioles and venules during exercise. Many substances produced during muscle contraction (e.g. potassium, adenosine, protons and P_i) can act as vasodilators.

The total peripheral extraction of oxygen from blood during exercise at $\dot{V}_{O_2max}$ in humans is approximately 80% and may be up to 95% across the active muscle mass (Fig. 3; [26]).

However, the extraction of blood-borne glucose and/or FFAs for oxidative phosphorylation in mitochondria tends to be much less. There are complex transmembrane transport processes that play a critical role in maintaining adequate flux rates of substrate delivery (see earlier).

Structural aspects

The capillarization and flow of red blood cells through muscle is extremely complex. However, it may be hypothesized that capillarity should be linked or matched to the mitochondrial volume, thereby matching supply to demand for oxygen and substrate. There is evidence that the surface available for gas exchange in capillaries is roughly matched to the absolute volume of mitochondria: Each millilitre of mitochondria is associated with approximately 10 km of capillary length, 0.14 m^2 of capillary surface area or 0.16 ml of capillary volume [23]. However, there is remarkable variation when comparing species of different body mass, activity level, and even before and after training or altitude exposure. This variation suggests that capillary network design is but one of many factors regulating tissue gas and substrate exchange. Clearly, capillaries must also serve other functions. In highly glycolytic muscle (with a very low mitochondrial volume) there is still a significant capillarization. It has been hypothesized that these capillaries are important for the removal of waste products, such as lactate, protons and heat [28].

Structural–functional interactions in supporting oxygen flow from red blood cells to mitochondria

The transfer of oxygen from the haemoglobin molecule into the mitochondria involves a series of resistive steps that must be overcome. The time available for exchange is a function of both capillary volume and blood flow. Estimates of capillary transit time range from 0.4 s to 0.8 s in exercising species, including humans, at $\dot{V}_{O_2max}$. This is an average estimate, since measurements of blood flow in individual microvascular units in conscious, exercising

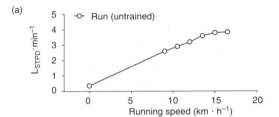

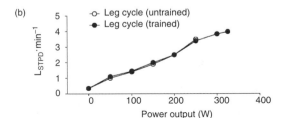

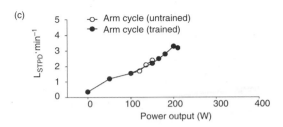

Fig. 4. Measurements of $\dot{V}_{O_2max}$ or 'peak $\dot{V}_{O_2max}$' during exercise

The attainment of a calculated $\dot{V}_{O_2max}$ in an untrained, young healthy male while running on a treadmill (a). $\dot{V}_{O_2}$ does not plateau when this same subject cycles with legs or arms, untrained or trained (b, c). Note that peak-$\dot{V}_{O_2}$ values measured during arm or leg cycling in the subject when untrained are only a fraction of $\dot{V}_{O_2max}$ but after endurance training these fractions can increase, and even surpass aerobic performance on the treadmill. This is an indication that training can be very specific for the muscle group and exercise mode used.

species are not available. Some microvascular units (the associated capillary network and mitochondrial volume) are likely to be underperfused, others are likely to be overperfused. It is not known whether the matching of heterogeneity of blood flow and oxidative metabolism measured in different muscle groups and muscle fibres is mere coincidence or indicates a fuctional link.

Recent evidence from human subjects exercising at $\dot{V}_{O_2max}$ suggest that the resistance to tissue gas exchange can play a role in partially limiting $\dot{V}_{O_2max}$ [29,30]. Most of the resistance, functionally measured as a large gradient in the partial pressure of oxygen between two compartments separated by a diffusion barrier, seems to occur between the red blood cell and the inner surface of the sarcolemma [29]. The lack of impediment in oxygen flow (measured as a low oxygen partial pressure gradient) across the muscle fibre may be due to facultative diffusion of oxygen bound reversibly to myoglobin. The extremely low partial pressure of oxygen (assumed to be almost zero) in the mitochondria will always allow an adequate oxygen flow whenever oxygen is present.

> ▶ During exercise, oxygen and substrates are delivered to active muscle fibres by the cardiovascular system — the degree of response is determined by the mass of muscle and the intensity of exercise.
> ▶ Active muscles release a variety of vasoactive agents during contraction, including potassium, adenosine, protons and P_i, all of which may mediate a local vasodilation.
> ▶ The surface available for delivery of oxygen from capillaries to muscle is roughly proportional to the volume of mitochondria.
> ▶ As well as delivering oxygen to muscle, capillaries are also important in the removal of waste products.

How is aerobic metabolism measured and how is it related to muscle activity?

The average daily metabolic rate (ADMR) measured with $^2H_2^{18}O$ rarely exceeds about twice the resting or basal level in sedentary, healthy human subjects. Even competitive, elite athletes in the Tour de France cycling championships can only maintain an ADMR between 3 and 4 times resting levels and this for only some 18–21 days [31].

One of the most widely accepted indices for the overall level of an individual's aerobic metabolism from lung to respiratory chain unit, during rest or exercise, is the rate at which oxygen is utilized or taken up at the whole-body level. This is referred to as whole-body oxygen consumption ($\dot{V}_{O_2}$, with units of volume per unit time and often per kg of body mass). This index, which is measured by indirect calorimetry, can be related directly to the amount of heat liberated from the oxidation of a known amount of carbohydrate or fat. Both $\dot{V}_{O_2}$ and heat production can be expressed in common units of energy expenditure (calories), but $\dot{V}_{O_2}$ is much easier to measure in exercising humans. All that is needed is a steady-state measure of the volume of air (or gas mixture) expired over a unit time ($\dot{V}_E$) using a pneumotachograph, collection bags, turbine or doppler ultrasound equipment and measurements of the concentrations or fractions of average inspired (F_IO_2) and expired (F_EO_2) oxygen. These measures can then be used to calculate $\dot{V}_{O_2}$:

$$\dot{V}_{O_2} = \dot{V}_E \cdot (F_IO_2 - F_EO_2) \tag{5}$$

During a graded, incremental exercise test on either a treadmill or cycle ergometer, whole-body $\dot{V}_{O_2}$ increases linearly with increasing intensity of exercise until a point is reached where further increases in the level of exercise do not evoke a proportional increase in $\dot{V}_{O_2}$ (Fig. 4). $\dot{V}_{O_2}$ is judged to have reached a plateau with increasing workloads and this is commonly referred to as $\dot{V}_{O_2max}$ and taken to indicate the maximal aerobic capacity. Higher workloads requiring greater ATP hydrolysis can only be sustained for a limited time by employing ever-increasing levels of anaerobic metabolism and PCr degradation. $\dot{V}_{O_2max}$ is regarded as an excellent index of the aerobic capacity of an individual with respect to a population, degree of physical training and during and after disease states. There are instances, however, when a true plateau in $\dot{V}_{O_2}$ does not occur [32]. Many physically untrained subjects, patients or subjects unfamiliar with ergometric testing do not exhibit a 'true' $\dot{V}_{O_2max}$ plateau, even though other commonly measured variables, such as heart rate and blood lactate concentration, may indicate that the subject is indeed performing to his/her maximal capability (Fig. 4). Furthermore, when exercising with smaller muscle masses, such as arm cranking, the highest $\dot{V}_{O_2}$ measured is but a fraction of whole-body exercise $\dot{V}_{O_2max}$ measured during treadmill running (Fig. 4). To circumvent this problem of detailed definition, 'peak-$\dot{V}_{O_2}$' has been adopted as the premier descriptor for levels of $\dot{V}_{O_2}$ that have been measured in circumstances where it is found that a true $\dot{V}_{O_2max}$ can not be measured.

As $\dot{V}_{O_2}$ appears to increase linearly with increasing work rates during submaximal exercise, it is possible to establish an index of efficiency of energy transduction. The choice of the correct parameters from which to calculate energy-transduction efficiency has been controversial. Gross mechanical efficiency (i.e. work performed/energy expended) is probably the easiest and most widely used and ranges from 15 to 25% for various types of exercise by different muscle groups at a range of intensities, the remainder of the energy consumption being lost or stored as heat. Other indices of 'efficiency' take into account resting metabolic energy expenditure, volume of active muscle mass and speed of movement, for example, but still efficiency is rarely greater than 30–35% [33].

In theory, during steady-state aerobic exercise, whole-body $\dot{V}_{O_2}$ represents the rate of ATP hydrolysis at the cross-bridge and thus, when referred to the substrate being oxidized, can be a relatively precise estimate of muscular energy consumption. Strictly speaking, this is an oversimplification, in as much as it does not account for changing levels of aerobic metabolism elsewhere in the body, such as the heart, respiratory muscles or visceral organs. It is also not possible to localize the consumption of oxygen within muscle — and any contribution from non-oxidative metabolic pathways, such as PCr/ATP breakdown or anaerobic glycolysis, are neglected. Recently, more refined techniques have been developed to calculate actual $\dot{V}_{O_2}$ within exercising muscle masses, such as

the knee extensor muscle group and the upper and lower arm, during submaximal and maximal dynamic and isometric contractions [26, 34,35]. The technique has as its core principle the relationship between $\dot{V}_{O_2}$ and cardiovascular oxygen transport (i.e. the Fick equation):

$$\dot{V}_{O_2} = \dot{Q}\,(C_{aO_2} - C_{vO_2}) \tag{6}$$

where $\dot{Q}$ is blood flow measured by thermodilution, dye dilution or Doppler ultrasound; C_{aO_2} is arterial oxygen concentration and C_{vO_2} is mixed effluent vein oxygen concentration. One important finding has been that during submaximal and maximal dynamic exercise (cycloergometry) the changes in $\dot{V}_{O_2}$ of the legs and whole-body $\dot{V}_{O_2}$ are similar and so changes in whole-body $\dot{V}_{O_2}$ may indeed be used as a relatively good indirect estimate of changes in $\dot{V}_{O_2}$ of an active muscle mass (Fig. 3; [26]).

The $\dot{V}_{O_2}$ measured across an active limb is still too global a measurement if cellular provision of ATP and mitochondrial function in different muscle fibres are to be studied in humans. For this level, oxygen consumption can be measured in suspensions of biopsied human muscle material in vitro [36]. More non-invasive techniques are being developed that make use of ^{31}P-MRS (magnetic resonance spectroscopy) to study certain aspects of human muscle aerobic function during submaximal and maximal muscle contraction patterns [37]. The relationship between changes in high-energy phosphate compounds, such as ATP, PCr and P_i, and pH and different levels of $\dot{V}_{O_2}$ have shown that there appears to be a tight coupling between submaximal aerobic metabolism and the degree of PCr degradation during exercise [37]. How this coupling or respiratory control is achieved is unknown at present, but it may operate through regulators such as [ADP], [P_i], [NADH] or Mi-CK [38].

▶ The rate of oxygen utilization at the whole-body level ($\dot{V}_{O_2}$) is widely used as an indicator of aerobic metabolism.
▶ Maximal aerobic capacity ($\dot{V}_{O_2max}$) is used as an index of the aerobic capacity of an indi-

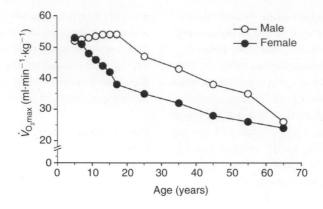

Fig. 5. The change in $\dot{V}_{O_2max}$ during the lifespan of males and females

vidual with respect to population, degree of physical training, etc.

▶ Recent techniques have enabled $\dot{V}_{O_2}$ to be calculated within exercising muscle masses.

▶ During maximal and submaximal dynamic exercise, changes in $\dot{V}_{O_2}$ of legs and the whole body are similar.

The adaptability of aerobic metabolism

It is not easy to bring together into one model the inter-relationships between ATP utilization at the cross-bridge, intracellular PCr shuttling, oxidative phosphorylation in the mitochondria, substrate fluxes into the TCA cycle from within and without the muscle fibre, peripheral oxygen and substrate delivery and finally oxygen consumption measured at the level of the exercising limb or whole body. Recent models describing the limits to aerobic capacity (i.e. $\dot{V}_{O_2max}$) stress that adaptability may occur at multiple sites ranging from the pulmonary gas-exchange system to the enzyme complement of the inner mitochondrial matrix [39,40].

Basic modelling of aerobic metabolism in exercising humans

$\dot{V}_{O_2max}$ is the best global index of aerobic capacity during exercise in an individual. It represents the maximal flux of oxygen from atmos-

phere to respiratory-chain units on the inner mitochondrial membrane. It does not account for the transfer of energy via the PCr shuttle to the site of ATP utilization at the actomyosin cross-bridge (see Fig. 1). Nevertheless, with certain caveats (see earlier), during steady-state exercise it has been assumed that $\dot{V}_{O_2}$ is equivalent to ATP utilization [41]:

$$\Delta ATP = \Delta \dot{V}_{O_2} + \Delta st O_2 + \Delta[LAC] + \Delta[PCr] \quad (7)$$

where $\Delta st O_2$ (muscle oxygen store), $\Delta[LAC]$ (lactate accumulation due to anaerobic metabolism) and $\Delta[PCr]$ (high-energy phosphate degradation) all equal zero. Recent evidence from studies using muscle biopsies and ^{31}P-MRS to measure high-energy phosphate status during submaximal exercise have shown that there is a graded reduction in PCr proportional to exercise work rate. The $\Delta[PCr]$ and $\Delta[LAC]$ may act as error signals for an adequate physiological response to exercise or simply by the consequence of Ca^{2+} activation and/or increased ATP turnover (see Table 1 and Fig. 2). High plasma [LAC] can inhibit adipose lipolysis and may play a role in controlling the fuel mix at higher exercise intensities. Infra-red spectroscopy technology has shown that myoglobin can be at least partially saturated during high-intensity aerobic exercise, indicating that $\Delta st O_2$ is probably minimal during steady-state exercise. As a gross approximation then, $\Delta \dot{V}_{O_2}$

probably does represent both oxygen flux to mitochondria and also an appropriate index of ATP flux (via the PCr shuttle) from mitochondria to actomyosin cross-bridge.

What then are the main determinants of $\dot{V}_{O_2max}$ in humans? The plurality of the question amplifies the fact that many aspects of aerobic metabolism may exert some restraint on $\dot{V}_{O_2max}$. The best-developed models which describe polyfactorial restraints on $\dot{V}_{O_2max}$ are simplistic at present. One model represents the flow of oxygen from atmosphere to the mitochondria as a cascade of resistive steps in series, nominally from alveolus to circulation and from circulation to muscle cells and finally mitochondrial oxygen utilization [39]. All these resistance steps, and any other restraints that may be operative (PCr shuttle efficiency, ATP hydrolysis efficiency), will have changing importance in setting aerobic exercise capacity in different scenarios. One important caveat exists when trying to assign importance, however: if one resistance step is experimentally changed, it is essential to monitor how all other resistance steps also change before assigning importance to the chosen variable. This is difficult in studies of exercising humans; however, there are some instances when one resistance can be changed when all (it is assumed) others are unchanged. For example, the addition of red blood cells to the circulation increases cardiovascular oxygen delivery overnight with no apparent change in mitochondrial enzyme activity or lung gas exchange. In this relatively straightforward example, approximately 70% of the total resistance to oxygen flow is calculated to reside within the cardiovascular system [30,39]. Where the other 30% of the restraint on $\dot{V}_{O_2max}$ resides is not known for certain, but it is thought, in normoxic conditions, to be downstream of the lung. Current research points to either tissue diffusion of oxygen, or its utilization in mitochondria, or both [26,29,30,42]. Further progress awaits the integration of several technologies to monitor changes in different segments of pathways involved in ATP production and utilization in muscle. The principle of multiple site control, here applied to $\dot{V}_{O_2max}$ limitation, has also been applied to metabolic control where a number of enzymes may all exert fractional constraint over the overall flow of energy through a particular pathway [43]. As yet, multifactorial control models have not been applied to $\dot{V}_{O_2max}$ and enzyme complement together.

Adaptability of aerobic metabolism to different conditions

In the following section, data from cross-sectional and longitudinal studies on changes in aerobic capacity are presented in the context of the multifactorial models described in the last section. The basic comparisons will include $\dot{V}_{O_2max}$ and skeletal muscle aerobic capacity during development/aging and after endurance training. Skeletal muscle aerobic capacity is the product of both muscle-mass-specific changes and changes in gross muscle mass.

Development and aging

Exercise tests measuring $\dot{V}_{O_2max}$ or peak $\dot{V}_{O_2}$ have been employed in studies of humans ranging from 5-year-old subjects to those in their nineties. There are large changes in body-mass-specific $\dot{V}_{O_2max}$ during development and aging in healthy, sedentary males and females (Fig. 5). Between the ages of 5 and 17 years, boys maintain a relatively high $\dot{V}_{O_2max}$ (approx. 45 ml·min^{-1}·kg^{-1}), whereas girls show a gradual decline from this initial value (-1.2 ml·min^{-1}·kg^{-1}·year^{-1} [44]). This difference between developing boys and girls is probably due to an underlying differential development in skeletal muscle mass (boys > girls) and adipose tissue (girls > boys) during adolescence. There are no major muscle-mass-specific changes in mitochondrial volume density, enzyme activities, capillarity or fibre-type proportion after the first year of life [45,46].

Between the 3rd and 7th decades, there is a gradual, steady rate of loss in $\dot{V}_{O_2max}$ of around -0.6 ml·min^{-1}·kg^{-1}·year^{-1} (males) and -0.3 ml·min^{-1}·kg^{-1}·year^{-1} (females). The decline in $\dot{V}_{O_2max}$ is less in females than in males. Beyond the age of 70 years, the rate of decline in $\dot{V}_{O_2max}$ may even increase. As with developing muscle masses in young and adolescent subjects, the muscle-mass-specific characteristics appear to remain relatively stable throughout life [47,48]. Once again, it is skeletal muscle mass that seems to be the variable most affected with age and, therefore, responsible for the decrease in $\dot{V}_{O_2max}$ [49]. Computed tomography scans of individual muscles show a progressive decrease in muscle cross-sectional area after the age of 30 [50], due to a selective loss of muscle fibres, but also to a decrease in muscle-fibre cross-sectional area [51]. There may also be ultrastructural disorganization of the sarcomere, dilatation of the sarcotubular system and decreased and thickened synaptic folds of motor endplates, leading to alterations in Ca^{2+} and ionic homeostasis in muscle.

One aspect of muscle-mass-specific aerobic capacity that may change with aging is mitochondrial respiratory-chain function. Whereas capillarity and the mitochondrial TCA enzyme pattern are relatively unaffected by age, recent evidence suggests that the respiratory rate of human muscle biopsy homogenates decreases significantly with age, coupled with a decrease in the enzyme activity of respiratory-chain complexes such as cytochrome oxidase [36]. No formal models have been designed to address changes in skeletal muscle or cardiovascular and respiratory function during healthy aging.

Endurance training

$\dot{V}_{O_2max}$ is increased with endurance training, the increase being proportional to the interaction of training intensity, frequency and total duration. Cross-sectional data on healthy, human subjects suggest that $\dot{V}_{O_2max}$ and muscle aerobic capacity, simply measured as mitochondrial volume or aerobic enzyme concentration [22], are related in a linear fashion, suggesting a general, genetically set, design principle [52]. However, when $\dot{V}_{O_2max}$ increases by 20% after endurance training, muscle aerobic capacity commonly increases by 40–100%. The adaptations in the trained muscle may actually be more closely related to changing local microconditions of oxygen and substrate supply and utilization during submaximal exercise, while changes in $\dot{V}_{O_2max}$ may be more closely related to changes in more global variables, such as

maximal mechanical power output [53]. The rates of changes in $\dot{V}_{O_2max}$, mitochondrial enzymes and substrate fluxes also show different time courses during training [54]. In general, changes in substrate fluxes and mitochondrial adaptations occur more quickly than those in $\dot{V}_{O_2max}$, suggesting either that there is greater resistance to change upstream of the muscle fibre in the oxygen-transfer cascade or that the changes in substrate utilization and mitochondrial aerobic capacity and $\dot{V}_{O_2max}$ are not causally related.

The specific adaptations within muscle aerobic machinery are extremely well documented, and refined models of muscle adaptation to long-term training or chronic stimulation in rodents — and, more recently, in humans — have been developed to study the mechanism of adaptation and its genetic expression [55]. The interplay between different substrate pathways after training is also well documented. However, the precise mechanism whereby mobilization and utilization of FFAs is increased, thereby sparing carbohydrate breakdown during submaximal exercise, is still a matter of intense controversy [13,15,53]. Similar disagreement surrounds the implications of changing the ratios of enzymes in mitochondria involved in different substrate pathways. The impact of changing enzyme ratios on the balance between proton production and ATP synthesis, and the consequent effects for the efficiency of cellular energy turnover, has yet to be fully explained.

If mitochondrial capacity for utilization of oxygen and substrates is increased during training, then it can be expected that there will be changes in local microvasculature as well. Structurally, capillarity is difficult to quantify precisely, because of its complex architecture. Difficulty in assessing patterns of perfusion, gas and substrate exchange *in vivo* also hinder precise measurements of capillary function and for that matter arteriolar and venular changes after endurance training. Nevertheless, gross indices of capillarity indicate that there is a training-induced increase in the capillary bed (and surface area), arteriolar dimensions and increased transport capacity for substrates [28].

Physical training can enhance $\dot{V}_{O_2max}$ in all age groups with similar benefits for young and old alike [44]. Skeletal muscle adaptations to endurance training are similar in young and old. Master athletes appear to maintain muscle oxidative capacity in spite of increased age, provided that they maintain an adequate level of endurance training [56]. Endurance training therefore appears to be a potent stimulus for increases in muscle aerobic capacity and $\dot{V}_{O_2max}$ at all ages. The interaction between training/detraining and human development, maturity and aging is not simple to model. Once again, there have been no formal attempts to dissect out the multifactorial limitations to $\dot{V}_{O_2max}$ after endurance exercise training.

▶ Skeletal muscle function and $\dot{V}_{O_2max}$ may change in several circumstances, such as during development and aging, and after endurance training.
▶ In most circumstances, skeletal muscle mass is significantly altered and has been linked to the change in $\dot{V}_{O_2max}$.
▶ The rates of adaptation are not similar for $\dot{V}_{O_2max}$, cardiorespiratory function and skeletal muscle metabolism.
▶ No formal attempts have been made to model multifuctional limitations to $\dot{V}_{O_2max}$ with healthy aging.

Conclusions

▶ The best index of overall aerobic capacity is whole-body $\dot{V}_{O_2max}$. However, there is a complex biochemical and topological arrangement of both structural and functional segments of the pathways involved in setting the maximal level of aerobic metabolism.
▶ High rates of delivery of oxygen and extramuscular substrates to the mitochondria are necessary for maximal cellular replenishment of intracellular PCr.
▶ The PCr shuttle acts as the essential link between ATP utilization, by contractile proteins and ionic pumps, and ATP production in the mitochondria.

▶ Modelling of overall oxygen transport from atmosphere to mitochondria and of metabolic pathways indicates that there are multiple sites of control or constraint, all of which have a fractional role to play in limiting the total flow of oxygen or substrate.
▶ The challenge now is to include the immense database on $\dot{V}_{O_2max}$, cardiorespiratory function and muscular aerobic capacity in humans of different ages, physical condition and in different environments into existing models so that the benefits of exercise in health and disease can be better evaluated.

Further reading
Crystal, R.G. and West, J.B., ed. (1991) *The Lung: Scientific Foundations* Raven Press, New York
Harden, E.S. and Terjung, R.L., ed. (1988) *Exercise, Nutrition and Energy Metabolism* MacMillan Publishing Company, New York
Hoppeler, H. and Billeter, R. (1991) Conditions for oxygen and substrate transport in muscles in exercising mammals. *J. Exp. Biol.* 160, 263–283
Poortmans, J.R., ed. (1993) *Principles of Exercise Biochemistry* Karger, Basel
Rowell, L.B. and Shepard, J.T., eds. (1996) *Handbook of Physiology: Section 12, Exercise* Oxford University Press, Oxford
Taylor, A.W., Gollnick, P.D., Green, H.J., Ianuzzo, C.D., Noble, E.G., Metivier, G. and Sutton, J.R., eds. (1990) *Biochemistry of Exercise, part VII* Human Kinetics Books, Champaign

References
1. Rall, J.A. and Wahr, P.A. (1993) Molecular aspects of muscular contraction. *Med. Sport Sci.* 38, 1–24
2. Hoppeler, H. and Billiter, R. (1991) Conditions for oxygen and substrate transport in muscles in exercising mammals. *J. Exp. Biol.* 160, 263–283
3. Sahlin, K. and Katz, A (1993) Adenine nucleotide metabolism. *Med. Sport Sci.* 38, 137–157
4. Katz, A. and Sahlin, K. (1988) Regulation of lactic acid production during exercise. *J. Appl. Physiol.* 65, 509–518
5. Weibel, E.R. (1984) *The Pathway for Oxygen: Structure and Function in the Mammalian Respiratory System.* Harvard University Press, Cambridge, MA
6. Greenhaff, P.L., Hultman, F. and Harris, R.C. (1993) Carbohydrate metabolism. *Med. Sport Sci.* 38, 89–136
7. Saltin, B. and Gollnick, P. D. (1988) Fuel for muscular exercise: role of carbohydrate. In *Exercise, Nutrition and Energy Metabolism* (Horton, E.S. and Terjung, R.L., eds.), pp. 45–71, MacMillan Publishing Company, New York
8. Saltin, B. and Gollnick, P.D. (1983) Skeletal muscle adaptability: significance for metabolism and performance. In *Handbook of Physiology: Section 10, Skeletal Muscle* (Peachy, L.D. ed.), pp. 555–631, American Physiological Society, Bethesda

9. Kayar, S.R., Hoppeler, H., Howald, H., Claassen, H. and Oberholzer, F. (1986) Acute effects of endurance exercise on mitochondrial distribution and skeletal muscle morphology. *Eur. J. Appl. Physiol.* **54**, 578–584

10. Bonen, A., McDermott, J.C. and Tan, M.H. (1990) Glucose transport in skeletal muscle. In *Biochemistry of Exercise, part VII* (Taylor, A.W., Gollnick, P.D., Green , H.J., Ianuzzo, C.D., Noble, E.G., Metivier, G. and Sutton, J.R., eds.), pp. 295–317, Human Kinetics Books, Champaign

11. Davies, C.T.M. and Thompson, M.W. (1979) Aerobic performance of female marathon and male ultramarathon athletes. *Eur. J. Appl. Physiol.* **41**, 233–245

12. Hoppeler, H. (1986) Exercise-induced ultrastructural changes in skeletal muscle. *Int. J. Sports Med.* **7**, 187–204

13. Holloszy, J.O. (1990) Utilization of fatty acids during exercise. In *Biochemistry of Exercise, part VII* (Taylor, A.W., Gollnick, P.D., Green, H.J., Ianuzzo, C.D., Noble, E.G., Metivier, G. and Sutton, J.R., eds.), pp. 319–327, Human Kinetics Books, Champaign

14. Srere, P.A. (1987) Complexes of sequential metabolic enzymes. *Annu. Rev. Biochem.* **56**, 89–124

15. Coggan, A.R., Spina, R.J., Kohrt, W.M. and Holloszy, J.O. (1993) Effect of prolonged exercise on muscle citrate concentration before and after endurance training in men. *Am. J. Physiol.* **264**, E215–E220

16. Tullson, P.C. and Terjung, R.L. (1991) Adenine nucleotide metabolism in contracting skeletal muscle. *Exercise Sports Sci. Rev.* **19**, 507–537

17. Sahlin, K., Katz, A. and Broberg, S. (1990) Tricarboxylic acid cycle intermediates in human muscle during prolonged exercise. *Am. J. Physiol.* **259**, C834–C841

18. Senior, A.E. (1988) ATP synthesis by oxidative phosphorylation. *Physiol. Rev.* **68**, 177–231

19. Schwerzmann, K. Hoppeler, H., Kayar, S.R. and Weibel, E.R. (1989) Oxidative capacity of muscle and mitochondria: correlation of physiological, biochemical and morphometric characteristics. *Proc. Natl. Acad. Sci. U.S.A.* **86**, 1583–1587

20. Cerretelli, P. and di Prampero, P. E. (1987) Gas exchange in exercise. In *Handbook of Physiology: Section 3, Respiratory System, 4* (Farhi, L.E. and Tenney, S.M., eds.), pp. 297–339, American Physiological Society, Bethesda

21. Dudley, G.A., Tullson, P.C. and Terjung, R.L. (1987) Influence of mitochondrial content on the sensitivity of respiratory control. *J. Biol. Chem.* **262**, 9109–9114

22. Hoppeler, H., Kayar, S., Claassen, H., Uhlmann, E. and Karas, K. (1987) Adaptive variation in the mammalian respiratory system in relation to energetic demand: III. Skeletal muscles; setting the demand for oxygen. *Respir. Physiol.* **69**, 27–46

23. Hoppeler, H., Mathieu-Costello, O, and Kayar, S.R. (1991) Mitochondria and microvascular design. In *The Lung: Scientific Foundations* (Crystal, R.G. and West, J.B., eds.), pp. 1467–1477, Raven Press, New York

24. Cogswell, A.M., Stevens, R.J. and Hood, D. (1993) Properties of skeletal muscle mitochondria isolated from subsarcolemmal and intermyofibrillar regions. *Am. J. Physiol.* **264**, C383–C389

25. Lewis, S.F., Taylor, W.F., Graham, R.M., Pettinger, W.A., Schutte, J.E. and Blomqvist, C.G. (1983) Cardiovascular responses to exercise as functions of absolute and relative work load. *J. Appl. Physiol.* **54**, 1314–1323

26. Knight, D.R., Poole, D.C., Schaffartzik, W., Guy, H.J., Prediletto, R., Hogan, M.C. and Wagner, P.D. (1992) Relationship between body and leg $\dot{V}_{O_2}$ during maxi-mal cycle ergometry. *J. Appl. Physiol.* **73**, 1114–1121

27. Eriksen, M., Waaler, B.A., Walloe, L. and Wesche, J. (1990) Dynamics and dimensions of cardiac output changes in humans at the onset and at the end of moderate rhythmic exercise. *J. Physiol. (London)* **426**, 423–437

28. Hudlicka, O., Brown, M. and Egginton, S. (1992) Angiogenesis in skeletal and cardiac muscle. *Physiol. Rev.* **72**, 369–417

29. Roca, J., Hogan, M.C., Story, D., Bebout, D.E., Haab, P., Gonzalez, R., Ueno, O. and Wagner, P.D. (1989) Evidence for tissue diffusion limitation of $\dot{V}_{O_2,max}$ in normal humans. *J. Appl. Physiol.* **67**, 291–299

30. Turner, D.L., Hoppeler, H., Noti, C., Gurtner, H.-P., Gerber, H., Schena, F., Kayser, B. and Ferretti, G. (1993) Limitations to $\dot{V}_{O_2,max}$ in humans after blood retransfusion. *Respir. Physiol.* **92**, 329–341

31. Westerterp, K.R., Saris, W.H.M., van Es, M. and ten Hoor, F. (1986) Use of the doubly labelled water technique in humans during heavy sustained exercise. *J. Appl. Physiol.* **61**, 2162–2167

32. Noakes, T.D. (1988) Implications of exercise testing for the prediction of athletic performance: A contemporary perspective. *Med. Sci. Sports Exercise* **20**, 319–330

33. Poole, D.C., Gaesser, G.A., Hogan, M.C., Knight, D.R. and Wagner, P.D. (1992) Pulmonary and leg $\dot{V}_{O_2}$ during submaximal exercise: Implications for muscular efficiency. *J. Appl. Physiol.* **72**, 805–810

34. Andersen, P. and Saltin, B. (1985) Maximal perfusion of skeletal muscle in man. *J. Physiol. (London)* **366**, 233–249

35. Ahlborg, G. and Jensen-Urstad, M. (1991) Arm blood flow at rest and during arm exercise. *J. Appl. Physiol.* **70**, 928–933

36. Trounce, I., Byrne, E. and Marzuki, S. (1989) Decline in skeletal muscle mitochondrial respiratory chain function: Possible factor in ageing. *Lancet* **i**, 637–639

37. Binzoni, T., Ferretti, G., Schenker, K. and Cerretelli, P. (1992) Phosphocreatine hydrolysis by ^{31}P-NMR at the onset of constant load exercise in humans. *J. Appl. Physiol.* **73**, 1644–1649

38. Krisanda, J.M., Moreland, T.S. and Kushmerick, M.J. (1988) ATP supply and demand during exercise. In *Exercise, Nutrition and Energy Metabolism* (Horton, E.S. and Terjung, R.L., eds.), pp. 27–44, MacMillan Publishing Company, New York

39. di Prampero, P.E. and Ferretti, G. (1990) Factors limiting maximal oxygen consumption in humans. *Respir. Physiol.* **80**, 113–128

40. Jones, J.H. and Lindstedt, S.L. (1993) Limits to maximal performance. *Annu. Rev. Physiol.* **55**, 547–569

41. di Prampero, P. E. (1981) Energetics of muscular exercise. *Rev. Physiol. Biochem. Pharmacol.* **89**, 144–222

42. Roca, J., Agusti, A.G.N., Alonso, A., Poole, D.C., Viegas, C., Barbera, J.A., Rodriguez-Riosin, R., Ferrer, A. and Wagner, P.D. (1992) Effects of training on muscle O_2 transport at $\dot{V}_{O_2,max}$. *J. Appl. Physiol.* **73**, 1067–1076

43. Kacser, H. and Porteus, J.W. (1987) Control of metabolism: What do we have to measure? *Trends Biochem. Sci.* **12**, 5–14

44. Krahenbuhl, G.S., Skinner, J.S. and Kohrt, W.M. (1985) Developmental aspects of maximal aerobic power in children. *Exercise Sports Sci. Rev.* **13**, 503–538

45. Bell, R.D., MacDougall, J.D., Billiter, R. and Howald, H. (1980) Muscle fiber types and morphometric analysis of skeletal muscle in six-year-old children. *Med. Sci. Sports Exercise* **12**, 28–31

46. Eriksson, B.O. (1980) Muscle metabolism in children: a review. *Acta Paediatr. Scand. Suppl.* **283**, 20–27

47. Grimby, G. (1990) Muscle changes and trainability in the elderly. *Top. Geriatr. Rehabil.* **5**, 54–62

48. Essen-Gustavsson, B. and Borges, O. (1986) Histochemical and metabolic characteristics of human skeletal muscle in relation to age. *Acta Physiol. Scand.* **126**, 107–114

49. Fleg, J.L. and Lakatta, E.G. (1988) Role of muscle loss in the age-related reduction in $\dot{V}_{O_2max}$. *J. Appl. Physiol.* **65**, 1147–1151

50. Rice, C.L., Cunningham, D.A., Paterson, D.H. and Lefcoe, M.S. (1989) Arm and leg composition determined by computed tomography in young and elderly men. *Clin. Physiol.* **9**, 207–220

51. Lexell, J., Taylor, C.C. and Sjostrom, M. (1988) What is the cause of the ageing atrophy? *J. Neurol. Sci.* **84**, 275–294

52. Bouchard, C., Chagnon, M., Thibault, M.-C., Boulay, M.R., Marcotte, M., Cote, C. and Simoneau, J.-A. (1989) Muscle genetic variants and relationship with performance and trainability. *Med. Sci. Sports Exercise* **21**, 71–77

53. Holloszy, J.O. and Coyle, E.F. (1984) Adaptations of skeletal muscle to endurance exercise and their metabolic consequences. *J. Appl. Physiol.* **56**, 831–838

54. Green, H.J., Helyer, R., Ball-Burnett, M., Kowalchuk, Symon, S. and Farrance, B. (1992) Metabolic adaptations to training precede changes in muscle mitochondrial capacity. *J. Appl. Physiol.* **72**, 484–491

55. Pette, D. and Vrbova, G. (1992) Adaptation of mammalian muscle fibers to chronic electrical stimulation. *Rev. Physiol. Biochem. Pharmacol.* **120**, 116–202

56. Holloszy, J.O and Kohrt, W.M. (1995) Exercise. In *Handbook of Physiology, Section 11, Ageing* (Masoro, E.J., ed.), pp. 633–666, Oxford University Press, New York and Oxford

4

Skeletal muscle metabolism during high-intensity exercise in humans

M.E. Nevill*[1] and P.L. Greenhaff†

*Department of Physical Education and Sports Science, Loughborough University, Loughborough, Leicestershire LE11 3TU, U.K., and †School of Biomedical Sciences, University Medical School, Queen's Medical Centre, Nottingham NG7 2UH, U.K.

Introduction

In 1962, Bergstrom and colleagues re-introduced and developed the needle-biopsy technique for obtaining human muscle samples. Subsequently, new analytical methods enabled metabolite and substrate concentrations to be measured, first in homogenates of the biopsy, and later in single-fibre fragments isolated from the biopsy by microdissection.

This chapter reviews how these techniques, in combination with a variety of exercise models, have contributed to our understanding of the metabolic response to high-intensity exercise in humans.

Exercise models and analytical techniques for examining muscle metabolism during high-intensity exercise

Isometric contractions and electrical stimulation

Metabolic studies involving isometric contraction have usually involved the measurement of isometric force production by the knee extensors. This is achieved by having subjects flex their knee joint at 90° over the end of a bed/chair and measuring force production by means of a strap around the ankle attached to a fixed strain-gauge. Isometric contraction is often used in conjunction with electrical-stimulation techniques. In the case of the quadriceps femoris muscles, two large electrodes are placed

proximally and distally to the anterolateral aspect of the thigh. The underlying muscles can them be stimulated to contract and isometric force production measured. The advantages of this approach are as follows: it does not depend on voluntary effort, and thereby motivation–contraction duration, pattern and intensity can be fixed exactly; muscle blood flow can be varied from totally intact to totally occluded; and muscle-biopsy samples can be obtained immediately after, or even during, contraction.

Dynamic contractions
Sprint cycling

The metabolic response to high-intensity exercise can be measured after maximum sprints performed on a cycle ergometer. The values for work and power output can be determined by multiplying the pedal or flywheel revolutions by the resistive force, although account should also be taken of the work required to accelerate the flywheel from a stationary or a rolling start. A typical power profile from such a test is shown in Fig. 1. The advantage of these performance tests is that they simulate closely the type and frequency of contraction in cycling activities. However, problems with data interpretation can arise, since throughout exercise torque is being generated at pedal velocities that are constantly changing.

Sprint running

Performance during running has been monitored by the determination of power output during non-motorized treadmill sprinting. The instantaneous product of restraint force and belt velocity is used to determine the horizontal

---

[1] To whom correspondence should be addressed.

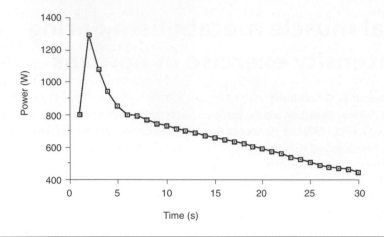

Fig. 1. Power output of one male subject during a 30 s maximal cycle sprint

component of power generated during the test. The advantages and disadvantages of this test are the same as those for sprint cycling, with the additional problem that several familiarization sessions are necessary before a subject can sprint maximally on the treadmill.

Isokinetic knee extension/cycling
Because sprint cycling and running involve accelerating and decelerating phases, the contracting muscles will be operating at an optimal velocity of shortening for maximum power for only short periods of time. It is possible, therefore, that problems may arise when attempting to relate the biochemical changes measured in biopsy samples to the observed decline in force or power production. To overcome some of these problems, isokinetic cycle ergometers and knee-extension dynamometers have been developed and applied to studies of human exercise performance.

The muscle-biopsy technique and tissue analysis
Muscle biopsies are taken under local anaesthesia using a stainless steel biopsy needle (4–6 mm diam.). The weight of the sample can be up to 200 mg (wet weight). If samples are to be used for metabolite analysis, the sample together with the needle is immediately plunged into liquid nitrogen to ensure rapid freezing. Samples are most frequently freeze-dried before analysis and, therefore, muscle metabolite values are expressed in terms of millimoles per kilogram of dry mass muscle [mmol·(kg of dm)$^{-1}$].

Analysis of metabolite concentrations in whole muscle
Although there are several methods for measuring metabolite concentrations, enzymic analyses are still the most commonly used because of their sensitivity and simplicity. For enzymic analysis of the homogenate of the whole biopsy, freeze-dried muscle biopsies are fat-extracted, dissected free of connective tissue, powdered and an acid extract obtained. The extract is neutralized and analysed for metabolites using spectrophotometric or fluorimetric techniques.

Analysis of metabolite concentrations in single fibres
Biopsies obtained from human muscle are composed of metabolically and functionally different fibre types. As the metabolic response to exercise differs markedly between fibre types, it may be important to be able to relate metabolic changes in the different fibre types to the development of fatigue in the whole muscle. Fragments of single muscle fibres can be dissected free from a muscle-biopsy sample with the aid of low-power microscopy. The ends of the fragments can then be used to identify the type

of fibre, and the remainder of the fibre fragments used for metabolite analysis using fluorimetric and luminometric techniques.

▶ Isometric contraction can be measured using electrical-stimulation techniques.
▶ Dynamic contraction can be determined by measuring work and power output of a subject cycling on an ergometer or sprinting on a treadmill.
▶ The muscle-biopsy technique combined with enzyme analysis, fluorimetry, luminometry and spectrophotometry is used to measure concentration of metabolites in whole muscle as well as in single fibres.

Requirements for ATP during intense contraction

Free energy can broadly be defined as the energy which is used to do work. In skeletal muscle the liberation of free energy, and thereby contraction, occurs solely as a result of the degradation of ATP to ADP and AMP. During contraction, ATP is required for a variety of cellular functions, including transport of calcium across the membrane of the sarcoplasmic reticulum and for electrolyte transport in the

sarcolemma and T-tubule system. The majority of ATP, however, is hydrolysed in the actomyosin ATPase reaction, which drives the contractile mechanism. It has been estimated that the ATP utilized in this ATPase reaction alone amounts to 8.6 mmol·(kg of dm)$^{-1}$·s^{-1} during the first few seconds of electrically induced isometric contractions at 50 Hz, and that the total ATP utilization including the sodium/potassium and calcium ATPases may be around 11 mmol·(kg of dm)$^{-1}$·s^{-1} (Table 1). Similar rates of ATP utilization of between 10 and 15 mmol·(kg of dm)$^{-1}$·s^{-1} have been reported for maximal sprint-cycling exercise [3]. Unfortunately, the muscle store of ATP is limited [~24 mmol·(kg of dm)$^{-1}$] and, therefore, the high utilization rates during exercise would, if unaccompanied by resynthesis, result in depletion of ATP stores in ~2 s. However, since muscle ATP content rarely falls by more than 30–40%, ATP must be resynthesized at a rate which almost matches the rate at which it is being depleted. It is possible to sustain a high rate of ATP utilization for more than 2 s because of the substantial contribution made by phosphocreatine (PCr) and glycogen degradation to ATP resynthesis. It is not possible, however, to maintain maximal anaerobic ATP turnover rates for long and, as Table 1 indicates, within 5 s of maximal isometric contraction

Table 1. Rates of anaerobic ATP resynthesis from phosphocreatine (PCr), glycolysis and PCr + glycolysis, during 30 s of near-maximal intensity, isometric muscle contraction in humans.

Duration (s)	ATP production rate [mmol·(kg of dm)$^{-1}$·s^{-1}]		
	PCr	Glycolysis	PCr + glycolysis
0–1.3	9.0	2.0	11.0
0–2.6	7.5	4.3	11.8
0–5.0	5.3	4.4	9.7
0–10	4.2	4.5	8.7
10–20	2.2	4.5	6.7
20–30	0.2	2.1	2.3

Values were calculated from metabolic changes measured in biopsies obtained during intermittent, electrically evoked contraction (1.6 s stimulation at 50 Hz, 1.6 s rest). Values calculated from [1,2].

total anaerobic ATP production has already begun to decline.

PCr degradation

Several reactions and metabolic pathways act in an integrated manner to meet the very high rates of ATP resynthesis required during intense exercise. Skeletal muscle contains a reservoir of high-energy phosphoryl groups in the form of PCr, amounting to 70–80 mmol·(kg of dm)$^{-1}$ at rest, which can be used to resynthesize ATP from ADP via the creatine kinase (CK) reaction. For each mole of PCr degraded, 1 mol of ATP is resynthesized:

$$PCr + ADP + H^+ \rightarrow creatine + ATP$$

During the early 1960s it was thought that the initial 10–15 s of maximal exercise was fuelled almost solely by PCr degradation. This was because PCr is stored in the cytosol in close proximity to the sites of energy, and because its hydrolysis does not depend on oxygen availability or necessitate the completion of several metabolic reactions before energy is liberated to fuel ATP resynthesis. While it is now accepted that glycolysis makes a significant contribution to ATP resynthesis during the initial seconds of maximal exercise (see next section), the importance of PCr still lies in the extremely rapid rates at which ATP can be provided.

The rapid reduction of PCr during intense exercise in man was first demonstrated experimentally by Hultman et al. [4], who showed that PCr degradation during dynamic cycling exercise was inversely related to absolute exercise intensity. In the 1980s, a number of studies confirmed that PCr was rapidly degraded during sprint exercise and that after only a few seconds of exercise the ATP yield from PCr was markedly reduced. Table 1 indicates that the rate of PCr hydrolysis during maximal exercise is at its highest at the beginning of the exercise. After only 2.6 s, the ATP yield from PCr is reduced by ~15%, and after 10 s of contraction it is reduced by more than 50%. The contribution of PCr to ATP resynthesis in the last 10 s of a 30 s bout of exercise is relatively small, amounting to 2% of the initial yield. Clearly, if

high-intensity exercise is to be continued, there must be a marked contribution to ATP resynthesis from other sources. The integrative nature of metabolism is illustrated by the fact that ADP, AMP, IMP (inosine monophosphate), NH$_3$ and P$_i$ (the products of ATP hydrolysis) are potent stimulators of glycogenolysis and glycolysis.

Glycogenolysis and glycolysis

Glycogenolysis is the hydrolysis of glycogen to glucose 6-phosphate. Glycolysis is the subsequent degradation of glucose 6-phosphate, derived principally from glycogen during maximal exercise, to the 3-carbon compound pyruvic acid. The generation of ATP in glycolysis occurs via the phosphorylation of ADP in the reactions catalysed by the enzymes phosphoglycerate kinase and pyruvate kinase. However, the reduction of NAD$^+$ in the glyceraldehyde 3-phosphate dehydrogenase reaction means that NADH must be reoxidized at the same rate as glycolysis, if glycolytic ATP production is to be maintained. The reoxidation of NADH may be accomplished by entry of electrons from NADH to the mitochondria via the malate–aspartate shuttle, or alternatively by the reaction catalysed by lactate dehydrogenase (LDH), which results in the formation of lactic acid and occurs in the cytoplasm. When the immediate fate of pyruvate is reduction to lactic acid, the pathway is sometimes referred to as 'anaerobic glycolysis' and may be summarized by the following reaction:

$$Glucose\ 6\text{-}phosphate + 3ADP + 3P_i$$
$$\rightarrow 3ATP + 2\ lactic\ acid = 2H^+$$

Thus for each mole of lactic acid produced, 1.5 moles of ATP are resynthesized and ATP yield can be calculated from the lactic aid produced. The substantial contribution that anaerobic glycolysis makes to ATP resynthesis during intense contraction is demonstrated by the very high muscle lactate concentrations [>100 mmol·(kg of dm)$^{-1}$] after maximal exercise. As Table 1 indicates, it seems that glycolysis is initiated almost immediately from the onset of exercise. After only 1.3 s of isometric contraction, the rate of ATP resynthesis from glycolysis was equal to 2.0 mmol·(kg of dm)$^{-1}$, and resulted in the formation of 1.7 mmol·(kg of dm)$^{-1}$ of lactate. It should also be noted that, unlike PCr degradation, glycolysis does not reach its maximal rate until after 5 s of exercise.

In summary, the ATP yield from anaerobic metabolism during intense contractions may be estimated by summing the changes in ATP, PCr and lactate, and multiplying by the appropriate ATP yield as described below.

ATP yield = [(change in ATP $\times$ 2)
  $-$ (change in ADP)]
  $+$ (change in PCr) + (change in lactate $\times$ 1.5)

Aerobic metabolism

Aerobic metabolism also contributes to ATP resynthesis during high-intensity dynamic exercise. Bangsbo and colleagues [5] showed that the contribution from aerobic and anaerobic sources to ATP resynthesis during the initial 30 s of 3 min exercise at 130% maximal oxygen utilization ($\dot{V}_{O_2,max}$) was 20% and 80% respectively. As exercise continued beyond 30 s, the contribution from aerobic energy production increased, whereas that from anaerobic sources declined correspondingly. Similarly, when repeated bouts of near-maximal exercise are interspersed with short periods of recovery, the contribution from aerobic energy production seems to increase with each successive bout of exercise.

> ▶ The high rates of ATP utilization during maximal sprint cycling would result in depletion of ATP stores within 2 s if not accompanied by resynthesis.
> ▶ A high rate of ATP utilization can be maintained for more than 2 s because PCr and glycogen degradation contribute to ATP synthesis.
> ▶ During intense exercise, PCr is used to resynthesize ATP from ADP very rapidly via the CK reaction, the rate of PCr degradation being inversely proportional to the intensity of exercise.

> ▶ During intense exercise, the glycolysis pathway generates ATP from the breakdown of glycogen (via glucose 6-phosphate and pyruvate).
> ▶ Aerobic metabolism makes a much smaller contribution to ATP generation during the first 30 s of high-intensity exercise than does anaerobic metabolism. But as exercise continues, the aerobic contribution increases.

Effect of exercise duration, intensity and type on muscle metabolism

Duration and intensity of exercise

As stated earlier, a high rate of anaerobic ATP resynthesis can only be maintained for short time-periods. Table 1 shows the rate of ATP resynthesis from muscle PCr and glycolysis over a period of 30 s at near-maximal isometric contraction in man. The rate of PCr degradation is at its maximum almost immediately and begins to decline after only 1.3 s of contraction. Conversely, the corresponding rate of glycolysis does not peak until after ~5 s of contraction and does not begin to decline until after 20 s of contraction. This suggests that the rapid utilization of PCr may buffer the momentary lag in energy provision from glycolysis, and that the contribution of the latter to ATP resynthesis rises as exercise duration increases and PCr availability declines. It is important to note that there is a progressive decline in the rate of ATP resynthesis from both substrates (and therefore total anaerobic ATP production) following their initial peaks. For example, during the last 10 s of exercise, the rate of ATP hydrolysis had declined to 2% (PCr) and 40% (glycogenolysis) of their respective maximal rates of ATP production. Similar responses have been shown to occur during isokinetic and dynamic exercise.

The mechanisms responsible for the almost instantaneous decline in the rate of PCr utilization during maximal exercise are unknown at present, but may be related to a decline in its availability. Considering the high energy demands of maximal exercise it is possible that the local availability of PCr at sites of rapid energy translocation within the cell (acto-myosin cross-bridges) could be responsible for the very rapid fall in PCr utilization at the onset of contraction. If intense exercise is continued for more than 20 s, the observed very low rate of PCr degradation can be attributed to its almost total depletion (Table 1).

The mechanisms responsible for the decline in glycogen utilization during intense exercise are unclear. However, it is unlikely that the depletion of muscle glycogen stores can be responsible, as levels have been shown to be high at the end of a bout of contraction which produced a 50% decline in glycogen utilization [1]. An exercise-induced, pH-mediated decrease in the activity of phosphorylase and phosphofructokinase (PFK) has been suggested as a mechanism which may be responsible for the decline in glycogen utilization during intense exercise. PFK catalyses the conversion of fructose 6-phosphate to fructose 1,6-bisphosphate and is the rate-limiting step in glycolysis. It is now generally accepted, however, that any potential for a pH-mediated inhibition of glycolysis during exercise is overcome in $vivo$ by the accumulation of allosteric activators of PFK (e.g. AMP and ammonia). Spriet and colleagues [6] showed that glycolysis can still be maintained in electrically stimulated human skeletal muscle despite a fall in muscle pH from 7.0 to ~6.4.

Both AMP and IMP have been associated with the regulation of glycogen degradation during exercise. Recent work demonstrated that a small increase in AMP concentration (10 μmol·l^{-1}) can increase the in $vitro$ activity of phosphorylase a. Furthermore, evidence demonstrating a close relationship between muscle ATP turnover and glycogen utilization in $vivo$ suggests that exercise-induced increases in AMP and P_i may be key regulators of glycogen degradation during muscle contraction [7]. Thus it is possible that, after the initial 20 s of maximal exercise when glycogen utilization is maximal, a decrease in the sarcoplasmic concentration of AMP (as a consequence of a decrease

in the rate of ADP formation and/or a pH-induced increase in the activity of AMP deaminase, the enzyme which catalyses the deamination of AMP to IMP) will result in a diminished activation of phosphorylase a and will thereby be responsible for the decline in glycogen utilization during maximal exercise.

Type of exercise: single versus repeated bouts of dynamic exercise

If a single bout of maximal exercise is repeated the metabolic responses to the first and subsequent bouts of exercise are different. For example, ten 6 s sprints were performed on a cycle ergometer [3]: during sprint one there was a 44 mmol·(kg of dm)$^{-1}$ decrease in PCr, a 3 mmol·(kg of dm)$^{-1}$ decrease in ATP and a 25 mmol·(kg of dm)$^{-1}$ increase in muscle lactate; during sprint ten, however, PCr decreased by only 25 mol·(kg of dm)$^{-1}$ and there was no decrease in ATP or increase in lactate. From these changes it was calculated that the rate of ATP resynthesis from anaerobic metabolism was 14.9 mmol·(kg of dm)$^{-1}$·s^{-1} in the first sprint, but only 5.4 mmol·(kg of dm)$^{-1}$·s^{-1} in the tenth. However, despite the rate of anaerobic energy production falling by more than 60% by sprint ten, power output was relatively well maintained at 73% of that observed in sprint one. The authors suggest this was due to a greater contribution from aerobic metabolism in the final sprint; however, even a vastly increased contribution from aerobic metabolism does not adequately explain the discrepancy between energy supply and performance. It is possible though that the contribution from PCr degradation was underestimated in the tenth sprint due to the need to biopsy a few seconds before the start of exercise and/or because some PCr resynthesis occurred immediately after the tenth bout of exercise before the final biopsy sample was obtained.

Nevertheless, it is clear that power output can be relatively well maintained in intermittent exercise and this is likely to be because the contribution from aerobic energy production during exercise is increased and/or the recovery periods between exercise allow a rapid resynthesis of PCr {approximately 1–3 mmol·(kg of dm)$^{-1}$·s^{-1}; [8,9]}. If the bouts of exercise are of a sufficiently short duration relative to the recovery periods, performance can be completely maintained for a limited period of time. When forty 15 m sprints were performed with a 30 s rest between each sprint, no difference was found between the time for the first and last sprints (2.63 s compared with 2.62 s); but if the same distance was covered as fifteen 40 m bouts, fatigue was evident in an increase in performance time from 5.61 s to 6.19 s [10].

▶ During maximal exercise, the contribution of glycolysis to ATP resynthesis rises with exercise duration, whereas that from PCr degradation declines.

▶ The rate of PCr utilization may decline due to a lack of availability of the substrate.

▶ After ~20 s of intense exercise, PCr stores are almost totally depleted.

▶ The decline in glycogen utilization after the initial 20 s of maximal exercise may be due to a decrease in [AMP] in the sarcoplasm, which diminishes phosphorylase a activation.

▶ Power output can be well maintained during bouts of exercise despite the decline in rate of anaerobic ATP resynthesis. This is probably due to the rapid resynthesis of PCr during recovery periods and/or an increase in aerobic energy production during exercise.

Effect of creatine supplementation on performance and metabolism during intense exercise

While the previously mentioned studies have suggested that PCr availability is critical for energy supply and performance during short-duration intense contraction, creatine (Cr) supplementation studies have examined this issue directly. Cr is a naturally occurring compound found principally in skeletal muscle and, in both free and phosphorylated forms, it appears to play a pivotal role in the regulation and

homoeostasis of skeletal-muscle energy metabolism and fatigue. The 'PCr/Cr shuttle' has been suggested to be a key element in mitochondrial/myofibrillar energy transfer and, therefore, in the control of mitochondrial respiration [11].

Endogenous synthesis of Cr occurs in liver, kidney and pancreas. In addition, dietary sources are found in meat and fish and it has been known for some time that oral ingestion of Cr will add to the whole-body Cr pool. It has been shown that ingestion of 20–30 g of Cr a day for 5–6 days can lead to more than a 20% increase in human muscle total Cr content, of which approximately 20–30% is in the form of PCr. It also appears that muscle Cr uptake is augmented if sub-maximal exercise is performed during the period of supplementation and that the extent of tissue uptake is dependent upon the initial tissue Cr content [12]. Further evidence suggests that Cr ingestion can increase significantly the amount of work which can be performed during repeated bouts of high-intensity exercise [13]. It has been proposed that these effects could have been due to an increase in muscle Cr content resulting in an accelerated PCr resynthesis in the 1 min recovery period between exercise bouts. This suggestion is supported by the lower accumulation of plasma ammonia during exercise after Cr ingestion which, under these conditions, is an accepted marker of net muscle ATP loss and cellular energy crisis. The positive effect of Cr ingestion on muscle PCr resynthesis during recovery from maximal exercise has also been confirmed [14].

▶ Creatine occurs naturally in skeletal muscle in both free and phosphorylated forms.

▶ Creatine, in its free and phosphorylated forms, plays a central role in energy transduction.

▶ A significant increase in the amount of work that can be performed during bouts of high-intensity exercise has been observed after supplementing diet with creatine for a few days before exercise.

Table 2. Muscle metabolites [mmol·(kg of dm)$^{-1}$, mean $\pm$ SEM, $n = 8$] at rest, after 0 s, 90 s, 3 min and 6 min of recovery from 30 s of cycle ergometer sprint exercise and percentage recovery (mean) in peak and mean power output

	Concentration of metabolites				
	Rest	0 s	90 s	3 min	6 min
Glycogen	321.5 $\pm$ 18.2	211.6 $\pm$ 18.5	223.2 $\pm$ 19.5	217.2 $\pm$ 21.0	221.0 $\pm$ 18.3
PCr	77.1 $\pm$ 2.4	15.1 $\pm$ 1.0	49.7 $\pm$ 1.1	57.2 $\pm$ 2.0	65.5 $\pm$ 2.2
ATP	25.6 $\pm$ 0.4	18.1 $\pm$ 1.7	19.1 $\pm$ 0.9	18.8 $\pm$ 1.1	19.5 $\pm$ 0.9
Lactate	3.8 $\pm$ 0.3	119.0 $\pm$ 4.6	107.3 $\pm$ 3.8	95.4 $\pm$ 5.6	81.9 $\pm$ 6.0

	Mean percentage recovery				
	Rest	0 s	90 s	3 min	6 min
Peak power output	–	–	77	86	87
Mean power output	–	–	74	84	90

Modified from [9].

Effect of training on muscle metabolism during intense exercise

As PCr availability appears to be critical for short-term performance, it might be expected that a possible training adaptation would be an increase in the muscle content of PCr at rest. However, this does not appear to be the case. High-intensity training improves performance during intense exercise, but the mechanism of adaptation is unclear. The training-induced improvements in sprint-running performance are associated with an increased contribution to energy supply from anaerobic glycolysis, as evidenced by a 20% increase in post-exercise muscle lactate content after training without any change in the resting muscle content of PCr and ATP [15]. This increased contribution from anaerobic glycolysis may be facilitated by an increase in PFK activity and an enhancement of skeletal muscle buffering capacity. There may also be an enhanced degradation of PCr in the first few seconds of a sprint after training which would be consistent with the finding of an increased CK activity.

Metabolism during recovery from intense exercise

In the mid-1970s, Harris and colleagues reported that in humans the time-course of PCr resynthesis after dynamic exercise or isometric contraction to exhaustion was biphasic, with an initial fast component having a half-time of 21–22 s and a much slower second component [8]. More recently, it has been shown in humans that the recovery of force after isometric knee extension exercise follows a similar time-course to the recovery of PCr, despite the persistence

Table 3. Resting PCr and glycogen contents [mmol·(kg of dm)$^{-1}$] and rates of degradation [mmol·(kg of dm)$^{-1}$·s^{-1}] in type-I and -II muscle fibres during 30 s of maximal treadmill sprinting and 30 s of intermittent electrical stimulation (1.6 s at 50 Hz stimulation, 1.6 s rest) with circulation occluded and intact

	Type-I		Type-II	
	PCr	Glycogen	PCr	Glycogen
Resting	79.4 ± 2.4	399 ± 28	89.6 ± 5.2	480 ± 33
30 s treadmill sprint	1.97 ± 0.01	2.57 ± 0.48	2.48 ± 0.08	4.21 ± 0.53
30 s electrical stimulation (occluded)	2.77 ± 0.10	2.05 ± 0.70	2.86 ± 0.17	4.32 ± 0.54
30 s electrical stimulation (intact)	1.63 ± 0.08	0.18 ± 0.14	1.87 ± 0.10	3.54 ± 0.53

Reproduced from [17–19].

of a high muscle lactate concentration and thus low calculated pH [16]. Such findings have begun to challenge the view held for many years that accumulation of lactic acid (particularly the hydrogen ion) is a key factor in causing the decline in performance or fatigue which is seen during intense exercise in humans.

The recovery of muscle metabolites and of power output after maximal cycling exercise has been examined by Bogdanis and colleagues [9]. Subjects performed two 30 s sprints on three occasions with varying rest intervals. A summary of the muscle metabolites after one 30 s sprint, and at 90 s, 3 min and 6 min of recovery is shown in Table 2. In early recovery, PCr increased rapidly from 20% of the resting value at the end of the sprint to 64% of rest at 90 s. During this time-period, muscle lactate only decreased from 119 to 107 mmol·(kg of dm)$^{-1}$. However, power was recovered to more than 70% of the value attained in the first sprint and a statistically significant relationship was found between peak power recovery and PCr resynthesis. The results of this study suggest that high power outputs can be attained in spite of a high muscle lactate and probably high muscle H$^+$ content. However, Sahlin and Ren [16], who came to similar conclusions in their study

examining maximal isometric contractions, point out that if the ability to sustain performance or isometric hold-time to exhaustion is examined, then accumulation of H$^+$ may limit ability to rephosphorylate ADP.

▶ High-intensity training improves performance during intense exercise due to an increased contribution to energy supply from anaerobic glycolysis and enhanced degradation of PCr (during the first few seconds of a sprint).
▶ Accumulation of H$^+$ ions is a key factor in the decline of performance and onset of fatigue during intense exercise.
▶ PCr is resynthesized rapidly during recovery periods (64% of resting value after 90 s recovery).

Muscle metabolism in single fibres during intense exercise

The conclusions presented so far have been based on metabolite changes measured in homogenates of biopsies obtained from vastus lateralis muscle. Human skeletal muscle, how-

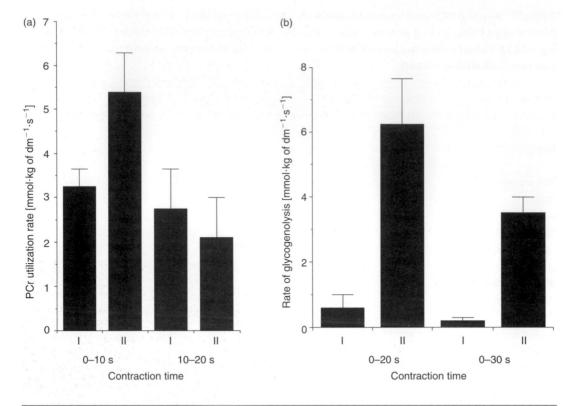

Fig. 2. Rates of PCr hydrolysis (a) and glycogenolysis (b) in type-I and -II muscle fibres during maximal, electrically evoked, intermittent (1.6 s stimulation at 50 Hz, 1.6 s rest) isometric contraction with circulation intact

(a) Rates of PCr utilization in type-I and -II muscle fibres during 0–10 s and 10–20 s of maximal isometric contraction. (b) Rates of glycogenolysis in type-I and -II muscle fibres during 0–20 s and 0–30 s of maximal isometric contraction. Values are given as means ± SEM.

ever, is composed of at least two major functionally and metabolically different fibre types. Clearly, it will be important to understand the significance of that fibre-type variability for a proper understanding of muscle function and fatigue.

Table 3 demonstrates that, at rest, PCr and glycogen contents are higher in type-II muscle fibres than in type-I fibres and, during intense contraction, the rates of glycogenolysis and PCr degradation are higher in type-II than in type-I fibres. This is true for both dynamic exercise (treadmill sprinting [17]) and electrically induced isometric contractions with circulation intact [18]. The rates of glycogenolysis recorded in both fibre types during treadmill sprinting and intermittent isometric contraction with circulation occluded is in good agree-

ment with the V_{max} of phosphorylase measured in both types, suggesting that glycogenolysis is occurring at a near-maximal rate during intense contraction. Surprisingly, during intermittent isometric contraction with circulation intact, the rate of glycogenolysis in type-I fibres is almost negligible, whereas the corresponding rate in type-II fibres is similar to that seen during contraction with circulatory occlusion. This suggests that, during maximal exercise, glycogenolysis in type-II fibres is invariably occurring at a maximal rate irrespective of the experimental conditions, whereas the rate in type-I fibres is probably very much related to tissue oxygen availability or the muscle duty cycle.

Further studies have attempted to relate the decline in whole-muscle isometric force production during intense, intermittent, electrically evoked contraction to the muscle changes occurring in individual muscle fibre types [20]. Fig. 2(a) shows the rates of PCr utilization in type-I and -II fibres between 0 and 10 s and 10–20 s of intense, electrically evoked contraction. During the first 10 s of stimulation the rates of utilization of type-I and -II fibres were 3.3 and 5.3 mmol·(kg of dm)$^{-1}$·s^{-1}, respectively. However, during the second period of stimulation, the rate of utilization in type-II fibres declined by ~60% to 2.1 mmol·(kg of dm)$^{-1}$·s^{-1} and, by the end of contraction, stores were nearly depleted, while the corresponding rate in type-I fibres remained relatively unchanged [2.8 mmol·(kg of dm)$^{-1}$·s^{-1}]. As already discussed, the rate of whole-muscle glycolysis does not begin to decline until after ~10 s of intense contraction (Table 1). Fig. 2(b) shows the rate of single-fibre glycogen degradation during 0–20 s of stimulation and 0–30 s of stimulation. During the initial 20 s of stimulation the rate of glycogenolysis in type-II fibres was rapid [6.3 mmol·(kg of dm)$^{-1}$·s^{-1}] compared with the negligible rate observed in type-I fibres [0.6 mmol·(kg of dm)$^{-1}$·s^{-1}], and was in excess of both the measured and calculated maximal rates of glycogen utilization determined for mixed-fibre muscle. When electrical stimulation was maintained for 30 s, similar to PCr the rate of glycogenolysis in type-II fibres declined by ~45% to 3.5 mmol·(kg of dm)$^{-1}$·s^{-1}, whereas the corresponding rate in type-I fibres remained very low. As expected during maximal exercise (and as previously described in dynamic exercise, Fig. 1), after the initial few seconds of contraction whole-muscle force production also declined by ~40% during the 30 s of stimulation. It would appear, therefore, that in parallel with the loss of force production there was a marked decline in the rates of PCr and glycogen utilization in type-II fibres. After 20 s of stimulation, the type-II fibre PCr store was almost totally depleted and the rate of glycogen utilization was decreasing. The consequent decrease in the rate of ATP resynthesis was thus insufficient to maintain force production, and fatigue occurred.

► At rest, levels of both PCr and glycogen are higher in type-II than in type-I muscle fibres.
► During both maximal dynamic exercise and isometric contraction, the rates of PCr degradation and glycogenolysis are higher in type-II than in type-I fibres.
► A fall in anaerobic ATP provision in type-II fibres seems to be associated with fatigue development during maximal exercise.

Conclusions

► During maximal exercise ATP is supplied by maximal rates of PCr degradation and glycolysis. As exercise progresses, ATP turnover is reduced due to lack of availability of PCr and reduced rate of glycolysis, resulting in a reduction in force and power.
► During intermittent exercise of short duration (6s or less), performance can be better maintained, probably as a result of adequate resynthesis of PCr in the intervening recovery periods.
► The importance of PCr availability to performance is confirmed by Cr-supplementation studies which show increased PCr content at rest and improved performance.
► It has been shown that force and power output recover rapidly after a single bout of maximal dynamic or isometric exercise, whereas muscle pH remains depressed.
► Single-fibre studies have shown that PCr and muscle glycogen contents are higher in type-II than in type-I fibres, and that during maximal exercise the rate of PCr and glycogen degradation are almost twice as high in type-II as in type-I fibres.
► Fatigue during high-intensity exercise in humans is due to the direct effect of H$^+$ on the contractile mechanism and a

59

reduction in energy supply resulting from lack of availability of PCr and reduction in the rate of glycolysis.

References

1. Hultman, E., Bergstrom, M., Spriet, L.L. and Soderlund, K. (1990) Energy metabolism and fatigue. In *Biochemistry of Exercise,* part VII (Taylor, A.W., ed.), vol. 21, pp. 73–92, Human Kinetics, Champaign

2. Hultman, E.A. and Sjoholm, H. (1983) Energy metabolism and contraction force of human skeletal muscle *in situ* during electrical stimulation. *J. Physiol.* 345, 525–532

3. Gaitanos, G.C., Williams, C., Boobis, L.H. and Brooks, S. (1993) Human muscle metabolism during intermittent exercise. *J. Appl. Physiol.* 75, 712–719

4. Hultman, E.A., Bergstrom, J. and McLennan-Anderson, N. (1967) Breakdown and resynthesis of phosphorylcreatine and adenosine-triphosphate in connection with muscular work in man. *Scand. J. Clin. Lab. Invest.* 19, 56–66

5. Bangsbo, J., Graham, T., Johansen, L., Strange, S., Christensen, C. and Saltin, B. (1992) Elevated muscle acidity and energy production during exhaustive exercise in man. *Am. J. Physiol.* 263, R891–R899

6. Spriet, L.L., Soderlund, K., Bergstrom, M. and Hultman, E. (1987) Anaerobic energy release in skeletal muscle during electrical stimulation in man. *J. Appl. Physiol.* 62, 611–615

7. Sahlin, K., Gorski, J. and Edstrom, K. (1990) Influence of ATP turnover and metabolite changes on IMP formation and glycolysis in rat skeletal muscle. *Am. J. Physiol.* 259, C409–C412

8. Harris, R.C., Edwards, R.H.T., Hultman, E., Nordesjo, L.O., Nylund, B. and Sahlin, K. (1976) The time course of phosphorylcreatine resynthesis during recovery of the quadriceps muscle in man. *Pflugers Arch.* 367, 137–142

9. Bogdanis, G.C., Nevill, M.E., Boobis, L.H., Lakomy, H.K.A. and Nevill A.M. (1993) Recovery of power output and muscle metabolites following 30 s of maximal sprint cycling in man. *J. Physiol.* 482, 467–480

10. Balsom, P.D., Seger, J.Y., Sjodin, B. and Ekblom, B. (1992) Physiological responses to maximal–intensity intermittent exercise. *Eur. J. Appl. Physiol.* 65, 144–149

11. Bessman, S.P. and Geiger, P.J. (1981) Transport of energy in muscle: The phosphorylcreatine shuttle. *Science* 211, 448–452

12. Harris, R.C., Soderlund, K. and Hultman, E. (1992) Elevation of creatine in resting and exercised muscle of normal subjects by creatine supplementation. *Clin. Sci.* 83, 367–374

13. Greenhaff, P.L., Casey, A., Short, A.H., Soderlund, K., Harris, R.C. and Hultman, E. (1993) Influence of oral creatine supplementation on muscle torque during repeated bouts of maximal voluntary exercise in man. *Clin. Sci.* 84, 565–571

14. Greenhaff, P.L., Bodin, K., Soderlund, K. and Hultman, E. (1994) Effect of oral creatine supplementation on skeletal muscle phosphocreatine resynthesis. *Am. J. Physiol.* 266, E725–E730

15. Nevill, M.E., Boobis. L.H., Brooks, S. and Williams, C. (1989) Effect of training on muscle metabolism during treadmill sprinting. *J. Appl. Physiol.* 67, 2376–2382

16. Sahlin, K. and Ren, J.-M. (1989) Relationship of contraction capacity to metabolic changes during recovery from fatiguing contraction. *J. Appl. Physiol.* 67, 648–654

17. Greenhaff, P.L., Nevill, M.E., Soderlund, K., Bodin, K., Boobis, L.H., Williams, C. and Hultman, E. (1994) The metabolic responses of human type I and II muscle fibres during maximal treadmill sprinting. *J. Physiol.* 478, 149–155

18. Greenhaff, P.L., Ren, J.-M., Soderlund, K. and Hultman, E. (1991) Energy metabolism in single human muscle fibres during contraction without and with epinephrine infusion. *Am. J. Physiol.* 260, E713–E718

19. Greenhaff, P.L., Soderlund, K., Ren, J. and Hultman, E. (1993) Energy metabolism in single human muscle fibres during intermittent contraction with occluded circulation. *J. Physiol.* 460, 443–453

20. Soderlund, K., Greenhaff, L.P. and Hultman, E. (1992) Energy metabolism in type I and II human muscle fibres during short term electrical stimulation at different frequencies. *Acta Physiol. Scand.* 144, 15–22

5

The balance of carbohydrate and lipid use during sustained exercise: the cross-over concept

George A. Brooks

Exercise Physiology Laboratory, Department of Integrative Biology, 5101 VLSB, University of California, Berkeley, CA 94720, U.S.A.

Introduction

Exercise intensity, training status, nutritional state and probably other factors interact in a complex manner to determine the fuel mix used during exercise. The cross-over concept provides a theoretical basis upon which to understand the interactions of these factors on the balance of carbohydrate and lipid metabolism during sustained exercise [1]. According to the cross-over concept, endurance training results in circulatory and muscular biochemical adaptations increasing lipid oxidation and decreasing the sympathetic nervous system (SNS) responses to given sub-maximal exercise stresses. These adaptations promote lipid oxidation during mild-to-moderate-intensity exercise. In contrast, increases in exercise intensity augment contraction-induced muscle glycogenolysis, alter the pattern of fibre-type recruitment and increase SNS activity. Therefore, the pattern of substrate utilization in an individual at any point in time depends upon the interaction between the exercise-intensity-induced responses (which increase carbohydrate utilization) and the endurance training-induced responses (which promote lipid oxidation).

The cross-over point is identified as the power output at which energy from carbohydrate-derived fuels equals energy from lipids, with further increases in power eliciting a relative increment in carbohydrate utilization and decrement in lipid oxidation. The contemporary literature contains data to indicate that, in post-absorptive individuals, mild-to-moderate-intensity exercises [$\leqslant$45% maximal oxygen consumption ($\dot{V}_{O_2max}$)] will be accomplished with lipid as the main substrate. There are also reports which indicate that, during high-intensity exercise (approx. 65% $\dot{V}_{O_2max}$), carbohydrate is the predominant substrate [2–5]. During training and competition, most athletes will perform at intensities which elicit more than 70–75% of maximum aerobic power and, as a result, they will be dependent upon carbohydrates for energy. Similarly, for non-athletes engaged in sustained, high-intensity recreational and occupational activities, carbohydrates are the main energy substrates. However, lipid is the predominant fuel during mild-to-moderate-intensity exercise as well as during recovery from prolonged exercises which result in glycogen depletion [1].

Classic measures utilizing indirect calorimetry

The cross-over concept is supported by several lines of evidence, including classic measurements of respiratory gas exchange as well as contemporary measurements of metabolite fluxes and oxidation involving muscle-biopsy procedure and isotopic tracers. Although contemporary biopsy and isotope measurements provide greater specificity, these results need to be interpreted within the context of total-body metabolism, as assessed by indirect calorimetry. Therefore, we will commence our assessment of fuel use during exercise with a discussion on the data obtained from indirect calorimetry.

Classically, measurements of the respiratory gas exchange ratio ($R = \dot{V}_{CO_2}/\dot{V}_{O_2}$; where $\dot{V}_{O_2}$ is oxygen consumption and $\dot{V}_{CO_2}$ is CO_2 output) have been used to estimate the mix of carbohydrate and lipid oxidized under diverse cir-

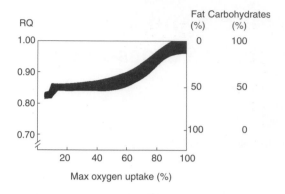

Fig. 1. Schematic illustration of the non-protein RQ at rest and during graded exercise
To the right, values are given as percentage contributions of fat and carbohydrate to energy yield. Reproduced from [2] with permission; data taken from [6] and other original sources.

cumstances. The history of studies on the relationships between exercise intensity and duration and the influences of diet is extensive, with some of the most important results attributable to the studies of Christensen and Hansen [6]. Nevertheless, it is well understood that using the respiratory gas exchange ratio (R) as an estimate of the cellular respiratory quotient (RQ) requires several assumptions. The main concern during exercise is that the use of bicarbonate to buffer organic acids (mainly lactic acid) produced during glycolysis will give rise to 'non-metabolic CO_2', thereby causing an overestimation of the CO_2 production from carbohydrate oxidation [7]. Therefore, investigators usually attempt to measure R during prolonged, steady-state exercise conditions when R and blood lactate concentration are both constant.

The extensive database available indicates that, at rest, for post-absorptive subjects who have been eating a standard mixed diet, R will approximate 0.83; put another way, the contributions of carbohydrate and fat to energy production are 44% and 56%, respectively. However, as illustrated by Åstrand and Rodahl [2], during exercise at even mild intensities, R easily reaches 0.85, indicating that approximately half of the energy is derived from carbohydrate, and

half from fat. Moreover, during moderate-intensity, steady-rate exercise, R approximates 0.90, indicating that energy from carbohydrate oxidation is predominant (68%). Thereafter, as R rises with the increase in power output, the contribution of carbohydrate fuel sources to energy production increases exponentially (Fig. 1).

Studies on marathon runners illustrate several physiological phenomena, including the mobilization and utilization of oxidizable energy sources. Although the marathon was once conceived as the longest running distance that a human could endure, athletes now regularly compete in longer runs. However, the marathon remains the longest Olympic running event and is an excellent experimental model to evaluate substrate utilization during prolonged, sub-maximal exercise. The measurements of O'Brien et al. [8] on post-absorptive men not given any calorific supplement (Fig. 2) indicate that marathon running is accomplished at a mean R of 0.95 (84% of energy derived from carbohydrate oxidation). These results indicate that the body's entire glycogen reserves are mobilized, probably by flux through the lactate pool, and so it remains that the marathon is the

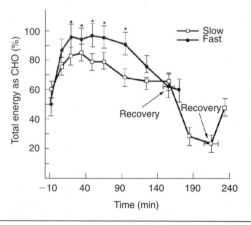

Fig. 2. Percentage of energy yield for carbohydrate before, during and after treadmill marathon calculated from respiratory gas exchange ratio (R) in slow (□) and fast (●) male groups
*Abbreviation used: CHO, carbohydrate. *, statistically different significance between groups. Modified from [8] with permission. ©1993, Lippincott, Williams & Wilkins.*

longest distance that can be managed mostly on the basis of endogenous carbohydrate fuel sources. Because most athletic endeavours are undertaken for lesser durations — but at greater relative intensities — than marathon running, the fuel mix used for most sports and athletic activities necessarily involves a greater percentage of carbohydrate and a smaller percentage of lipid than is used in the marathon.

In concluding this section, it should be acknowledged that data such as are presented in Fig. 1 and Fig. 2 are taken from individuals who have adapted to a mixed diet, but who have fasted for 8–12 h. The results would be different (i.e. R would be higher at a lower relative intensity) if subjects had recently consumed a carbohydrate meal. Similarly, if subjects were adapted for several weeks to a protein–fat (ketotic) diet, R would also be different (i.e. lower) at low to moderate intensities [9]. Thus it must be acknowledged that the balance of fuels used during exercise are influenced by immediate as well as long-term eating habits.

▶ Increased intensity during exercise promotes utilization of carbohydrate.
▶ Endurance training leads to circulatory and muscular adaptations which enhance lipid oxidation.
▶ The cross-over point is the level of power output at which energy generated from carbohydrates is equal to energy from lipids.
▶ Indirect calorimetry techniques indicate that, at rest, the contributions of carbohydrate and fat to energy production are 44% to 56%, respectively. During mild-intensity exercise, contributions are approximately equal. During moderate exercise, the contribution from carbohydrate becomes predominant (68%) and continues to rise as exercise intensity increases.
▶ During marathon running, the mean contribution of carbohydrate to energy is 84%, which represents mobilization of the body's entire glycogen store.
▶ Muscle glycogen use during exercise is dependent upon the relative intensity of the exercise.

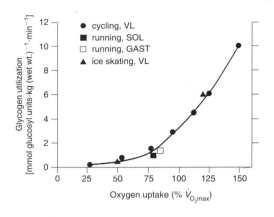

Fig. 3. Muscle glycogen utilization rates at various work intensities during different types of exercise
Abbreviations used: VL, vastus lateralis; SOL, soleus; GAST, gastrocnemius muscle. Reproduced from [10] with permission.

Muscle glycogen use during exercise

The conclusion that a progressive rise in carbohydrate utilization occurs with increases in exercise power output is supported by the extensive work of Hultman and associates using the muscle-biopsy technique. From the seminal studies of the late 1960s and early 1970s, Hultman and associates showed net muscle glycogen use to be highly dependent on relative exercise intensity as well as on the pre-exercise nutrition and activity habits which, in turn, affected the amount of muscle glycogen stored.

Hultman and Spriet [10] have reviewed the extensive database from biopsy studies relating exercise intensity and muscle glycogen catabolism in well-nourished subjects on a balanced (i.e. predominantly complex carbohydrate) diet. The data available reveal insignificant glycogen use at rest, little or no measurable net glycogen use during mild-intensity exercise and exponentially increasing glycogen use during moderate- and greater-intensity exercises (Fig. 3). The biopsy data as initially provided by Hultman and associates also make it clear that muscle glycogen is the major carbohydrate used to fuel short-term, as well as prolonged, bouts of exercise.

Tracer measurements of substrate oxidation

More recently, isotope technology has provided independent means to evaluate the effects of exercise intensity and training on blood glucose and plasma free fatty acid (FFA) fluxes. By this terminology, flux refers to the rate of entry (production) and exit (use) of metabolites in blood. An alternative term for entry is 'rate of appearance' (Ra), and for exit is 'rate of disappearance' (Rd). Because the liver is the main site for glucose production and entry into the blood, the Ra is considered to be due to the hepatic glucose production (HGP) [7]. Fortunately, there now exist several reports from independent laboratories to show that, as predicted from measurements of respiratory gas exchange and muscle glycogen content, blood glucose flux increases exponentially as a function of relative exercise effort, which is expressed as a percentage of $\dot{V}_{O_2max}$ [3].

Until recently, it was difficult to conceptualize results of studies measuring glucose and fatty acid fluxes. Many of the studies used too few subjects at too few exercise intensities. In addition, although it was clear that the blood glucose flux was proportional to exercise power output (i.e. glucose flux rose with increase in exercise power output), the same relationship was not readily observed between plasma FFA flux and exercise power output. Indeed, because it was generally considered that the supply of oxidizable fuel sources increased to match increases in metabolic rate, as given by the rate of oxygen consumption ($\dot{V}_{O_2}$), it was surprising that exercises which elicited 75% of $\dot{V}_{O_2max}$ were achieved without any increase in plasma FFA flux above rest.

Of the results now available from studies using indirect calorimetry as well as multiple isotopic tracers, the work of Romijn et al. [5] is representative. The data obtained on highly trained cyclists shows maximal plasma FFA utilization during relatively mild-intensity exercise (25% $\dot{V}_{O_2max}$), but diminished FFA use during hard-intensity exercise (85% $\dot{V}_{O_2max}$) (Fig. 4). Note also in Fig. 4 that the calculated use of other lipids (triacylglycerol) is greatest at the

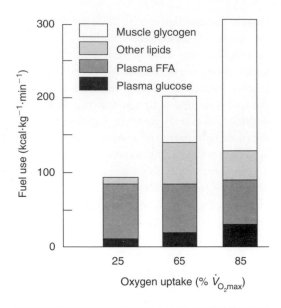

Fig. 4.   Illustration of fuel selection in highly trained male cyclists at three exercise intensities

Plasma glucose and calculated muscle glycogen use increase exponentially as exercise intensity increases. Conversely, plasma FFA use declines from a peak value at 25% of $\dot{V}_{O_2max}$. Modified from [5] with permission. ©1993, The American Physioligical Society.

moderate intensity (65% $\dot{V}_{O_2max}$). However, during exercise at high intensity, both FFA and other lipid use decline such that total lipid oxidation during exercise eliciting 85% of $\dot{V}_{O_2max}$ is no greater than during exercise at 25% of $\dot{V}_{O_2max}$.

The data of Romijn et al. are also illustrative of the roles played by glucose and other carbohydrate energy sources. As indicated already, glucose use scales exponentially to relative exercise power output. Similarly, muscle glycogen use scales exponentially to power output, but with a relatively greater gain in glycogen than in glucose use at high power output. Thus, during moderate- and greater-intensity exercises, muscle glycogen, not blood glucose, is the major carbohydrate for oxidation. Because the data represent precursor (starting glucose and glycogen), and final (end CO_2) points, the pathways of carbohydrate flux, involving lactate mainly, are not shown. The importance of flux through the lactate pool is discussed separately later.

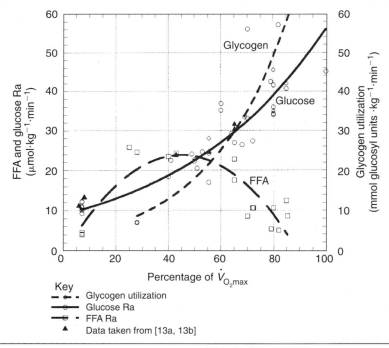

Key
- ● - Glycogen utilization
─○─ Glucose Ra
─□ - FFA Ra
▲ Data taken from [13a, 13b]

Fig. 5. Results of an extensive literature search showing blood glucose and FFA flux rates (Ra) and net muscle glycogenolysis [10] as functions of relative exercise intensity given as a percentage of $\dot{V}_{O_2max}$

This form of analysis indicates exponential increments in muscle glycogenolysis and glucose Ra as functions of relative exercise intensity. In contrast, the analysis shows multi-component polynomial response of plasma FFA flux, with mild-to-moderate-intensity exercise (i.e. 25–40% of $\dot{V}_{O_2max}$) eliciting a large rise in flux, but cross-over and decreasing flux at approximately 55% $\dot{V}_{O_2max}$. Note that plasma FFA flux is predicted to reach minimal values as $\dot{V}_{O_2max}$ is approached. Original glycogen-degra-dation data calculated per kg of active muscle, and nor-malized to 35 kg of active muscle to scale with glucose and FFA kinetics which are given as mg per kg of body weight per min. Redrawn from [3] with permission. ©1995, The American Physiological Society.

For glycogen utilization, $y = 2.11\ e^{(0.04x)}$, $r^2 = 0.87$; for glucose Ra, $y = 9.8\ e^{(0.02x)}$, $r^2 = 0.84$; and for FFA Ra, $y = -0.833 + 1.14x - 0.013x^2$, $r^2 = 0.87$.

Even though glucose use increases exponentially on an absolute basis as a function of exercise power output (relative to the increase in total carbohydrate use at high power outputs), the percentage of energy supplied by glucose decreases. For instance, in resting individuals, muscle glycogen is not used, and so glucose is the precursor for carbohydrate oxidation; as such, glucose provides close to 50% of the total energy flux. However, during exercise at 65% of $\dot{V}_{O_2max}$, glucose oxidation provides only approx. 15% of the energy flux even though the glucose flux has increased to about three times that at rest. This is because, relative to the increase in overall metabolic rate and carbohydrate oxidation, the increase in glucose flux and oxidation was small.

The fact that HGP and peripheral glucose uptake is limited to a 5–8-fold increase over resting levels during maximal-intensity exercise [11–13] is probably in the best interest of the organism [7]. If the capacity to increase muscle glucose uptake was much greater than the capacity to increase HGP (e.g. if it was possible to achieve during exercise a 20–30-fold increase in muscle glucose uptake over resting levels), then it is likely that exercise would result in profound hypoglycaemia. Therefore, we have evolved a system in which the switch to carbohydrate-based fuels during exercise occurs with reliance on muscle glycogen, not blood glucose (Fig. 5).

> ► Isotope technology enables the effect of exercise intensity and training on the blood glucose and plasma FFA fluxes to be determined.
> ► Blood glucose flux rises exponentially with increase in relative exercise effort (given as % $\dot{V}_{O_2max}$).
> ► FFA utilization is greatest during moderate-intensity exercise. At high intensities, use of FFA and triacylglycerol declines to the same levels as during mild-intensity exercise.
> ► During moderate- and greater-intensity exercise, muscle glycogen is the major carbohydrate for oxidation.
> ► At rest, blood glucose provides almost 50% of total energy flux, whereas at 65% $\dot{V}_{O_2max}$, it provides only ~15%.

The cross-over concept illustrated

Fig. 5 represents a compilation of studies available in the literature relating glycogen net change — and blood glucose and FFA fluxes — to relative exercise power output [3]. The line describing glycogen use is a segment of that from Hultman and Spriet [10] (Fig. 3). The lines describing glucose and FFA fluxes are computer-generated best fits of data compiled in the literature and include results from Romijn et al. [5]. Like glycogen use, glucose flux rises exponentially as a function of exercise power output. In contrast, FFA flux is described by an inverted hyperbola. Plasma FFA flux rises rapidly in the transition from mild-to-moderate-intensity exercise, with a peak at approx. 45% of $\dot{V}_{O_2max}$. However, in the transition from mild- to hard-intensity exercise (i.e. above approx. 65% $\dot{V}_{O_2max}$), FFA flux falls in absolute terms. Thus results obtained using muscle biopsies, indirect calorimetry and isotopic tracers indicate that, at high relative efforts, carbohydrate energy sources (not lipids) become the predominant substrate fuelling

muscular exercise. In other words, although lipids provide slightly more than half the energy at rest and during mild-intensity exercise, as exercise effort increases beyond moderate intensity, the relative contribution of fat decreases and that of carbohydrate increases; cross-over from lipid to carbohydrate dependence occurs.

Cellular control of substrate oxidation

Regulation of the balance of carbohydrate and lipid use in muscle cells during contractions is presently understood in terms of a combination of extracellular and intracellular factors. Extracellular factors include substrate delivery and neuroendocrine signals. Intracellular factors include decreases in the intramuscular ATP/ADP and NADH/NAD$^+$ ratios, as well as increases in cytosolic free calcium (Ca^{2+}), inorganic phosphate (P$_i$), cyclic AMP and malonyl coenzyme A (CoA) concentrations as relative effort increases [7,14–16].

Substrate delivery

Although generally overlooked, the delivery of fuels to working muscle during contractions is an important means of matching supply to demand. Because delivery of a metabolite to muscle depends on the product of blood flow and arterial concentration, both blood flow and metabolite concentration influence fuel supply and use.

In cardiovascular physiology, the redistribution (shunting) of blood flow to working muscle is an acknowledged means of increasing muscle oxygen supply during exercise. Similarly, the increases in cardiac output, fractional distribution of blood flow to the muscle and total muscle blood flow that occur during exercise represent important means to increase substrate delivery to contracting muscle [11–13]. Because absolute blood flow to some tissues and organs rises little during exercise, as compared with rest, redistribution of blood containing oxidizable substrates (glucose, lactate, FFAs) allows for preferential distribution of these metabolites to sites of oxidation in working muscle beds. Moreover, because arterial concentrations of metabolites can change significantly during exercise, and because muscle blood flow may increase as much as 20-fold (or more) during exercise, the vascular conductance of metabolites represents an extremely important means of controlling substrate utilization during exercise.

Glucose

If the arterial and venous blood metabolite contents across a tissue bed can be measured, the fractional extraction of that metabolite can be calculated. The fractional extraction of a metabolite equals $([v]-[a]/[a]) \times 100$; where $[v]$ and $[a]$ are venous and arterial concentrations, respectively. In addition, if tissue blood flow can be measured simultaneously, then the product of blood flow and arteriovenous metabolite concentration difference will yield the actual metabolite consumption across that tissue, organ or limb bed [11–13,17].

During a wide range of exercise intensities, arterial glucose concentration changes little in comparison with that at rest because, overall, HGP is well matched to removal. During short-term, high-intensity exercises, however, feed-forward regulation of HGP causes arterial glucose concentration to rise as production exceeds removal [12,13,17]. Conversely, during prolonged sub-maximal exercise, especially in the post-absorptive state, arterial glucose concentration can fall as HGP is limited by depletion of hepatic glycogen. Moreover, in prolonged exercise, depletion of skeletal muscle glycogen will result in decreased release of muscle lactate and a reduced supply of the precursor for hepatic gluconeogenesis. Thus muscle glycogen depletion will limit hepatic gluconeogenesis and glucose production [18].

For glucose, like other metabolites that exist in high concentrations, fractional extraction across a resting muscle bed is low (2–4%) [11,17]. Furthermore, glucose fractional extraction across muscle decreases during exercise. For example, in a resting post-absorptive person with an arterial glucose concentration of 4 mM, the femoral arteriovenous glucose con-

centration approximates 0.1 mM, making the fractional extraction 2.5%. However, during exercise at a power output eliciting 50% of $\dot{V}_{O_2max}$, glucose fractional extraction across the legs declines to approximately 0.05 mM, or approx. 1%. Thus, given the small arteriovenous glucose difference and low fractional extraction of glucose across muscle, blood flow becomes an important regulator of glucose metabolism.

FFAs

Relative to arterial glucose concentration in resting post-absorptive individuals (4–5 mM), the circulating FFA concentration is low, usually between 0.5 mM and 1.0 mM [11,17]. At present, it is difficult to know what the fractional extraction of fatty acids is in resting human muscle, because even the best arteriovenous FFA concentration measurements across limbs depend on the assay of venous blood, some of which has passed through adipose tissue. Thus the arteriovenous FFA difference across resting human limbs is negative (e.g. −0.005 mM), indicating lipolysis and FFA release into the central circulation. However, during moderate-intensity exercise (e.g. 50% $\dot{V}_{O_2max}$), blood flow through muscle increases much more than through limb adipose, the arteriovenous difference for FFA ($[a-v]_{FFA}$) better reflects muscle metabolism. In this case, $[a-v]_{FFA}$ becomes positive, and fractional extraction approximates 10% [17].

During prolonged, sub-maximal exercise lasting several hours, arterial FFA concentration can rise several-fold, so long as exercise intensity does not result in lactic acidosis. It has regularly been observed that lactic acidosis of hard exercise is accompanied by depressed circulating FFA concentrations, as well as decreased limb fractional uptake. The decline in circulating arterial FFA levels at power outputs greater than approximately 50% of $\dot{V}_{O_2max}$ can be attributed to a decline in adipose lipolysis, not increased uptake. Acidosis is suspected of stimulating adipocyte phosphodiesterase, thereby reducing adipocyte cyclic AMP concentration, hormone-stimulated lipolysis and FFA release from adipose tissue [7,19].

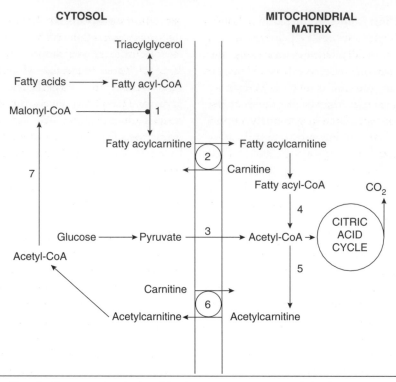

Fig. 6. Proposed relationship between pyruvate dehydrogenase, acetyl-CoA carboxylase and oxidative metabolism of fatty acids in the heart

1, carnitine palmitoyltransferase 1 (CPT1); 2, carnitine-acylcarnitine translocase; 3, pyruvate dehydrogenase; 4, β-oxidation; 5, carnitine acetyltransferase; 6, carnitine-acetylcarnitine translocase; 7, acetyl-CoA carboxylase. According to this scheme, increasing the glycolytic flux from glucose, or glycogen, will result in pyruvate formation, and the series of reactions identified as 3 → 6 → 7 → 1. Increased formation of malonyl CoA results in inhibition of CPT1, and decreased access of fatty acylcarnitine derivatives to the mitochondrial matrix. Reproduced from [16] with permission. ©1993, The American Society for Biochemistry and Molecular Biology.

Muscle GLUT-4 translocation

When contraction starts, GLUT-4 is translocated to the sarcolemma, an event permissive of increasing glucose uptake. Because muscle GLUT-4 translocation occurs in working muscle independent of changes in circulating insulin, there must be another signalling pathway to affect translocation. In addition to being involved in the excitation–contraction coupling process, phosphorylase activation and stimulation of the mitochondrial dehydrogenases, Ca^{2+} is also suspected of being involved in the GLUT-4 translocation signalling process [11,14]. These events, in combination with high V_{max} rates and abundance of glycolytic enzymes in muscle, give rise to rapid glycolysis leading to pyruvate and lactate formation. As previously mentioned, arterial FFA concentration falls during hard exercise, owing in part to the inhibition of lipolysis by lactic acidosis, and muscle hormone-stimulated lipase may be inhibited by the same mechanism. Thus carbohydrate is the predominant energy source during hard exercise, partly because acidosis limits the availability of circulating FFAs.

Muscle glycogenolysis and glycolysis

Muscle glycogenolysis during exercise is stimulated by Ca^{2+} as well as by adrenaline through a cyclic AMP-dependent mechanism. Of these two mechanisms, increased free Ca^{2+} is thought to be more important because glycogenolysis is activated very rapidly when contractions start, long before adrenaline levels change. However, adrenaline can amplify the rate of glycogenolysis in muscle, and can stimulate glycogenolysis and lactate release from inactive muscle and other tissues [14].

In skeletal muscle, glycolysis proceeds because of a complex set of interactions involving substrate availability, ATP/ADP and $NADH/NAD^+$ ratios, as well as levels of P_i and Ca^{2+}. Moreover, several of these (i.e. cytoplasmic redox or $NADH/NAD^+$) are influenced by mitochondrial function. (For a review of the factors regulating glycolysis in muscle, see [14].)

Pyruvate flux in muscle, either to lactate or to oxidation, is determined by pyruvate dehydrogenase. Pyruvate dehydrogenase is activated by rising Ca^{2+} and pyruvate concentrations as well as falling ATP/ADP and $NADH/NAD^+$ ratios [14,15]. Regardless of the specific extra- and intracellular signals governing glycogenolysis and glycolysis during exercise, the net effect will be an increased glycolytic carbon flux. In highly oxidative fibres, during mild-to-moderate-intensity exercise, the pyruvate flux equals or is less than the mitochondrial demand for oxidizable substrate. Therefore, the glycolytic flux is directed towards oxidation in mitochondria. However, during hard- and greater-intensity exercises, pyruvate production in excess of mitochondrial uptake results in lactate production even in fully aerobic red muscle fibres [14]. In fast-twitch, pale glycolytic fibres, glycolysis inevitably results in lactate formation due to imbalances in capacities for pyruvate production and oxidation [20–22].

Mitochondrial uptake of activated FFAs

As a mechanism to explain the observed interactions between carbohydrate and lipid oxidation, scientists are currently focusing on a mechanism involving the down-regulation of mitochondrial FFA uptake. Here, we present the model of Saddik et al. [16], which has been shown to operate in the heart. A surge in glycolytic flux in the cytosol serves to increase pyruvate production and transport into mitochon-

dria. In the mitochondrial matrix, an increased concentration of acetyl-CoA serves to form acetylcarnitine, which is exported to the cytosol. After conversion of acetylcarnitine to acetyl-CoA by acetyl-CoA carboxylase, cytosolic acetyl-CoA acts as a precursor to malonyl-CoA (Fig. 6). Malonyl-CoA inhibits mitochondrial carnitine palmitoyltransferase (CPT1), thereby limiting availability of activated fatty acids for transport into mitochondria for β-oxidation. Thus through a mechanism involving malonyl-CoA and CPT1, muscle contraction and hormonally mediated stimulation of glycogenolysis and glycolysis serve to decrease mitochondrial fatty acid uptake and the relative contributions of lipid to the fuel mix sustaining exercise at high power output.

Though appealing in many respects, the proposed mechanism of balancing the use of carbohydrate- and lipid-derived substrates is not without problem. The key issue is that in muscle and heart there always seems to exist more malonyl-CoA than is necessary to inhibit CPT1 completely, and block lipid oxidation. Therefore, malonyl-CoA must exist in intracellular compartments segregated from CPT1, or there must exist other, presently unknown, regulatory factors that permit mitochondrial uptake of activated fatty acids. Further, to date, it has not been possible to demonstrate a change in muscle malonyl-CoA during exercise.

Fibre-type recruitment

It is well established [20] that exercise is accomplished by an ordered recruitment of muscle fibres. According to the 'size principle', the recruitment of muscle fibres progresses from small motor units innervating slow oxidative fibres, to larger motor units involving fast-twitch fibres as power output and force requirements increase. Thus as exercise power output increases from mild to moderate, hard and maximal intensities, a greater muscle mass — including a greater percentage of fast-glycolytic muscle fibres — must be recruited. The increase in carbohydrate metabolism, and decrease in lipid metabolism, which occur at power outputs above 65% $\dot{V}_{O_2max}$ (Fig. 5) must represent, in part, the consequence of increased recruitment of fast-glycolytic muscle fibres. Unfortunately, at present it is not possible to quantify how much of the shift in substrate utilization is attributable to altered muscle recruitment.

Endocrine (sympathetic) signalling

Gradations in hormonal secretion appear gauged to alterations in exercise power output. In general, as exercise intensity increases, insulin decreases, whereas glucagon, adrenaline and noradrenaline increase. During exercise, changes in the individual concentrations of insulin and glucagon can vary little, but the insulin:glucagon ratio declines, signalling increased HGP. In concert, increases in circulating adrenaline and noradrenaline can influence HGP, particularly at the start of exercise, or during high-intensity exercise when the catecholamines signal skeletal muscle and hepatic glycogenolysis and gluconeogenesis [12,13].

▶ Like the control of breathing, which is acknowledged to represent a highly integrated and redundant system, the control of HGP during exercise is similarly complex.

▶ From the perspective of the cross-over concept, the role of the SNS seems particularly important [1].

▶ Feed-forward regulation of glycaemia, involving an exaggerated rise in HGP and a rise in arterial glucose concentration during high-intensity exercise, is associated with sympathetic stimulation [12,13].

▶ Sympathetic stimulation of muscle glycogenolysis can result in increased glycolytic flux, lactic acidosis and down-regulation of adipose lipolysis and mitochondrial FFA uptake [7].

Training increases the capacity to utilize lipids as energy sources for low- and moderate-intensity exercise

At least five types of finding support the idea that endurance training before sustained, low-to-moderate-intensity exercises of defined durations increases lipid oxidation. These are that training (i) increases the mass of the mitochon-

drial reticulum [23]; (ii) decreases the respiratory gas exchange ratio (R) [4]; (iii) spares muscle glycogen [24]; (iv) lowers circulating blood lactate [25]; and (v) lowers catecholamine levels within given exercise power outputs [26]. Because prior endurance training allows any given exercise task to be performed at a relatively lower intensity, these training-induced adaptations allow individuals to shift their relative position on the carbohydrate- and lipid-utilization curves (Fig. 5) and, thereby, to use relatively more lipid and less carbohydrate.

Training and the muscle mitochondrial reticulum

At the level of skeletal muscle, training increases mitochondrial mass [4,27,28] or, more specifically, the size of the mitochondrial reticulum [23], resulting in an increased capacity to oxidize all substrates, including fatty acids [4,29]. As recognized initially by Molé et al. [29], training increases the enzymes of FFA translocation, the tricarboxylic acid cycle, the β-oxidation pathway, and components of the electron-transport chain necessary to oxidize fatty acids. More recently, it has been recognized that increased mass of the mitochondrial reticulum due to training allows increased lipid oxidation and allows given rates of tissue oxygen consumption to be accomplished with higher [ATP]/[ADP], citrate and acetyl-CoA levels. The net result of these training-induced effects is superior cellular 'respiratory control', a down-regulation of phosphofructokinase and pyruvate dehydrogenase, decreased net glycolytic flux and increased lipid oxidation at a given metabolic flux [14,15,30]. In other words, increasing the mitochondrial mass and capacity to utilize activated fatty acids should promote activity of the glucose–fatty acid cycle of Randle et al. [15]. According to the cross-over concept, during sub-maximal exercise at a given, or greater, power output, the relative oxidation of lipid will increase, and carbohydrate will decrease after training.

Training and the respiratory gas exchange ratio (R)

The observation of a lower respiratory gas exchange ratio for exercise at a given sub-maximal power output after training is highly reproducible [4]. Although analysis of the effects of training on R is seldom accompanied by attempts to model changes of bicarbonate kinetics during exercise, use of 'steady-state' conditions with stable R values and blood lactate concentrations ensures that R approximates lumped average body respiratory quotient. The data obtained under steady-state conditions — showing a downward shift in R — probably represent the strongest data available to support the conclusion of enhanced lipid oxidation after training [7].

Muscle glycogen sparing

Biopsies of animal and human [20,24] limb skeletal muscle, before and after (absolute) exercise power output and duration, indicate that net muscle glycogen breakdown is less after training than before. This blunting of muscle glycogenolysis due to training is most prominent during the rest-to-exercise transition than later in exercise. Overall, a large volume of data indicates that training spares muscle glycogen during prolonged exercise [1,4].

Although the basic observation of glycogen sparing owing to prior exercise training is well documented, studies on the mechanisms explaining the lesser net change are lacking. For instance, glycogen synthesis has been observed during exercise [30a], but the phenomenon is little studied. Similarly, little attention has been given to evaluating the pathways of glycogen metabolism during exercise before and after training. The concept that glycogen sparing is associated exclusively with increased lipid metabolism has been challenged and it has been demonstrated that increased uptake of blood glucose explains most of the muscle glycogen sparing observed in trained rats during moderate to-hard-intensity exercise [31,32].

Blood lactate after endurance training

After training, blood lactate concentration is lower at a given sub-maximal exercise power output than before training [1,25,33]. In addition, after training the net change in active muscle glycogen during a given exercise task is less

than before training [4,33]. These results are usually taken as evidence that lipid energy sources substitute for carbohydrates (i.e. muscle glycogen and blood glucose) and, thereby, spare glycogen and decrease lactate production. A decrease in glycogenolysis during exercise of a given (or greater) power output due to reduced SNS stimulation after training favours decreased carbohydrate and increased lipid oxidation. However, this interpretation — that lower circulating lactate levels during exercise after training are due to lesser production — needs to be reconsidered in view of evidence, obtained using tracers and unlabelled lactate, which suggests that training increases lactate clearance during exercise [22,25].

Blood catecholamines after endurance training

Of the several endocrine factors which affect substrate utilization, adrenaline is probably one of the most important. In resting individuals, it is known that there is a hierarchy of metabolic effects of adrenaline [34], with lipolytic, glycogenolytic and insulin-suppressive effects occurring in sequence. Although the hierarchical effects of adrenaline are less-well documented during exercise than during rest, it is known that during exercise of mild-to-moderate intensity, circulating adrenaline levels rise little, if at all, compared with at rest [26]. Moreover, for a given exercise power output, circulating adrenaline is lower in trained than in untrained subjects. As a result, during prolonged low-to-moderate-intensity exercise, the lipolytic effects of adrenaline and other lipolytic hormones (e.g. growth hormone [35]) predominate in favour of lipid oxidation. However, at moderate-to-hard-intensity exercise in the trained state, SNS stimulation occurs and catecholamines rise [12,26,36], with the rise in catecholamines being greater in highly trained than in less-fit individuals [12]. Under these conditions, adrenaline stimulates muscle glycogenolysis. Lactate is produced as the result of mass action [14], and gluconeogenesis is facilitated by availability of a gluconeogenic precursor. At the same time, lactic acidosis acts to inhibit lipolysis [19]; noradrenaline rises, possi-

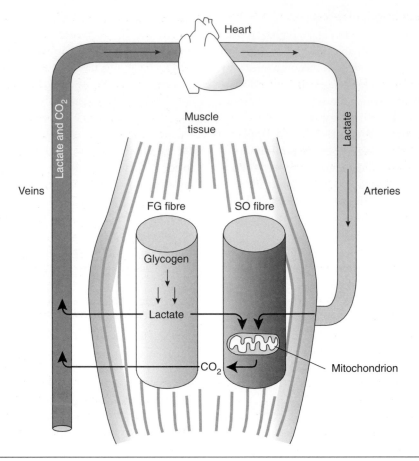

Fig. 7. Diagram of the lactate shuttle by which lactate formed in some cells and tissues can be oxidized in other cells and tissues

Abbreviations used: FG, fast, glycolytic; SO, slow, oxidative. Redrawn from Exercise Physiology: Human Bioenergetics and its Applications, Second Edition, by George A. Brooks, Thomas D. Fahey and Timothy P. White. Copyright ©1996 by Mayfield Publishing Company. Reprinted by permission of the publisher.

bly stimulating HGP [1,11–13]; and adrenaline suppresses insulin release, allowing the gluconeogenic effects of glucagon on hepatic function to predominate [12,13,17]. For these reasons, catecholamines probably play a powerful role in determining the cross-over point [1].

The view that there exist hormonal influences over the pattern of substrate utilization during exercise is supported by results recently provided by Green et al. [33]. They observed a significant change in the pattern of substrate utilization (i.e. lower lactate and R) after only a few exercise-training bouts, before circulatory or muscle mitochondrial adaptations occurred. Although Green et al. did not report cate-

cholamine responses during exercise before and after minimal training, they did succeed in uncoupling changes in patterns of substrate utilization from mitochondrial or cardiovascular adaptations. In terms of the cross-over concept, their results provide an opportunity for insight into metabolic regulation. If, with minimal training, there is insufficient time for mitochondrial or circulatory adaptations to occur, yet SNS activity is suppressed, the pattern of substrate utilization will nevertheless change — with cross-over occurring at a higher power output than before training. Thus with SNS down-regulation, more lipid will be used during moderate-intensity exercise even without the benefit of other physiological and biochemical adaptations.

The lactate shuttle during rest and exercise

One of the major recent conceptual advances in the fields of exercise physiology and intermediary metabolism concerns our understanding of lactic acid metabolism [21] (Fig. 7). In the past, lactic acid was considered to represent the product of oxygen-limited metabolism (anaerobic glycolysis) during exercise [37,38]. Lactic acid was viewed as a dead-end metabolite which accumulated during exercise, giving rise to fatigue and the oxygen debt [38]. In contrast, data have recently been obtained which have engendered the concept that the formation, exchange and utilization of lactic acid represent an important means of distributing carbohydrate energy sources after a carbohydrate meal [39] and during sustained physical exercise [11,-21,22,40–42]. In this scheme, lactic acid is an advantageous metabolic intermediate between carbohydrate storage forms (glucose and glycogen) and metabolic end-products (CO_2 and water). The advantage of lactic acid as a metabolic intermediate is that it exchanges rapidly between tissue compartments. Lactate is of low molecular mass, does not require insulin for transport and moves across cell-membrane barriers by facilitated transport [43]. Skeletal muscle, once considered to be the major site of lactic acid formation, may in some circumstances be responsible for significant net lactic acid removal from the blood [11]. The liver, once considered to be the major site of lactic acid removal through its participation in the Cori cycle may, in fact, contribute in a major way to the rise in arterial lactic acid concentration at the outset of strenuous exercise [22,41]. During exercise, when the arterial concentration rises, lactic acid may become the predominant fuel for the heart [44]. Other tissues and organs (e.g. skin and intestine) probably also contribute to the change in blood lactate concentration during exercise [22].

▶ Endurance training increases the mass of the mitochondrial reticulum, decreases R, spares muscle glycogen, lowers circulating blood lactate and lowers catecholamine levels, all of which lead to increased lipid oxidation during exercise.

▶ Lactic acid, which used to be viewed as a 'dead-end' by-product of exercise, is now considered to be an advantageous metabolic intermediate between carbohydrate storage forms and metabolic end-products, due to its rapid exchange between compartments.

▶ During exercise, most lactate (75–80%) is removed by oxidation but lactate conversion to glucose (20–25%) supports glucose homeostasis.

Applicability of the cross-over concept

Whether one evaluates the effect of exercise intensity on the balance of substrate utilization from measurements of respiratory gas exchange (Fig. 1), or other measures of metabolite fluxes (Fig. 5), the same conclusion is reached that the fuel mix switches (crosses over) from fat to carbohydrate in the transition from mild- to hard-intensity exercises. This means that, during exercises at which most sportsmen and -women train and compete, carbohydrate energy fuels predominate. Therefore, the movement patterns and recruitment of fibre types and metabolic systems that emphasize carbohydrate metabolism need to be emphasized in training [7].

In contrast, if the specific goal of exercise is to oxidize lipid stores, then prolonged mild-to-moderate-intensity needs to be prescribed ($\leqslant 50\%$ $\dot{V}_{O_2max}$). In this way, with a slightly lower intensity but several times greater duration, more total calories and a greater percentage of lipid can be utilized than in shorter-duration, higher-intensity exercise bouts.

In the absence of the technical capacities to infuse isotopes and take blood as well as muscle-biopsy samples, the inevitable question arises concerning a practical method to identify the cross-over point. One possible solution may lie in determination of the 'lactate threshold' (L_T).

L_T has engendered controversies concerning its detection and interpretation. Some of the proposed causes for the exponential rise in blood lactate at a power output eliciting approximately 65% of $\dot{V}_{O_2max}$ include: an imbalance between blood lactate appearance (Ra) and removal (Rd) in which Ra suddenly predominates; an adrenaline-stimulated increase in muscle glycogenolysis; and an increased recruitment of fast-glycolytic (pale) muscle fibres [22]. In addition, the presence of muscle O_2 insufficiency, and a Pasteur effect in the stimulation of glycolysis have been hypothesized. However, regardless of the exact individual or combination of factors giving rise to lactate acidosis and the accompanying increase in pulmonary minute ventilation, one common interpretation is an acceleration in glycogenolysis and glycolysis. Therefore, L_T (which can be detected from finger stick, ear lobe, or other blood), or the ventilatory threshold (V_T, which can be detected non-invasively from pulmonary gas exchange measurements) may offer opportunities to detect the cross-over point in post-absorptive individuals engaged in progressive-intensity exercise protocols. However, caution must be used when assessing the cross-over point from L_T or V_T measurements because of the indirect nature of those events as measurements of muscle metabolism. For instance, in an individual who has recently consumed a carbohydrate meal, the resting R as well as the average cellular respiratory quotient may easily exceed 0.85. In such cases, carbohydrate would be the predominant fuel at rest, a circumstance not detectable from abrupt changes in blood lactate or pulmonary minute ventilation during exercise [1].

Conclusions

▶ Numerous factors including exercise intensity, training status, nutritional state and probably other factors (e.g. gender) interact in a complex manner to determine the fuel mix used during exercise.

▶ The cross-over concept (Fig. 5) provides a theoretical basis upon which to understand the interactions of these factors on the balance of carbohydrate and lipid metabolism during sustained exercise.

▶ According to the cross-over concept, endurance training results in circulatory and muscular biochemical adaptations which enhance lipid oxidation as well as decrease the SNS responses to given sub-maximal exercise stresses. These adaptations promote lipid oxidation during mild-to-moderate-intensity exercise.

▶ Increases in exercise intensity augment contraction-induced muscle glycogenolysis, alter the pattern of fibre-type recruitment and increase SNS activity.

▶ The pattern of substrate utilization in an individual at any point in time depends upon the interaction between the exercise-intensity-induced responses (which increase carbohydrate utilization) and the endurance-training-induced responses (which promote lipid oxidation).

▶ Regardless of training state, during moderate- and higher-intensity exercises, carbohydrate energy sources (glycogen, lactate, glucose) will be the major energy sources.

References

1. Brooks, G.A. and Mercier, J. (1994) The balance of carbohydrate and lipid utilization during exercise: the crossover concept. *J. Appl. Physiol.* **76**, 2253–2261
2. Åstrand, P.-O. and Rodahl, K. (1970) *Textbook of Work Physiology*, p. 460, McGraw-Hill, New York
3. Brooks, G.A. and Trimmer, J. (1995) Literature data support the Crossover Concept. *J. Appl. Physiol.* **80**, 1073–1074
4. Gollnick, P.D. (1985) Metabolism of substrates: energy substrate metabolism during exercise and as modified by training. *Fed. Proc.* **44**, 353–357
5. Romijn, J.A., Coyle, E.F., Sidossis, L.S., Gastaldelli, A., Horowitz, J.F., Endert, E. and Wolfe, R.R. (1993) Regulation of endogenous fat and carbohydrate metabolism in relation to exercise intensity and duration. *Am. J. Physiol.* **265**, E380–E391
6. Christensen, E.H. and Hansen, O. (1939) Arbeitsfähigkeit und Ehrnährung. *Skand. Arch. Physiol.* **81**, 160
7. Brooks, G.A., Fahey, T.D. and White, T.P. (1996) *Exercise Physiology: Human Bioenergetics and its Applications*. Mayfield, Mountain View
8. O'Brien, M.J., Viguie, C.A., Mazzeo, R.S. and Brooks, G.A. (1993) Carbohydrate dependence during marathon running. *Med. Sci. Sports Exercise* **25**, 1009–1017

9. Phinney, S.D., Bistrian, B.R., Evans, W.J., Gervino, E. and Blackman, G.L. (1983) The human metabolic response to ketosis without caloric restriction: preservation of submaximal exercise capacity with reduced carbohydrate oxidation. *Metabolism* **32**, 769–776

10. Hultman, E. and Spriet, L.L. (1988) Dietary intake prior to and during exercise. In *Exercise, Nutrition and Energy Metabolism* (Horton, E.S. and Terjung, R.L., eds.), pp. 132–149, Macmillan, New York

11. Brooks, G.A., Wolfel, E.E., Groves, B.M., Bender, P.R., Butterfield, G.E., Cymerman, A., Mazzeo, R.S., Sutton, J.R., Wolfe, R.R. and Reeves, J.T. (1992) Muscle accounts for glucose disposal but not blood lactate appearance during exercise after acclimatization to 4,300m. *J. Appl. Physiol.* **72**, 2435–2445

12. Kjaer, M., Farrell, P.A., Chistensen, N.J. and Galbo, H. (1986) Increased epinephrine response and inaccurate glucoregulation in exercising athletes. *J. Appl. Physiol.* **61**, 1693–1700

13. Richter, E.A., Kiens, B., Saltin, B., Christensen, N.J. and Savard, G. (1988) Skeletal muscle glucose uptake during dynamic exercise in humans: role of muscle mass. *Am. J. Physiol.* **254**, E555–E561

13a. Friedlander, A.L., Casazza, G.A., Horning, M.A., Budinger, T.F. and Brooks, G.A. (1998) Effects of exercise intensity and training on lipid metabolism in young women. *Am. J. Physiol.* **275**, E853–E863

13b. Friedlander, A.L., Casazza, G.A., Horning, M.A. and Brooks, G.A. (1999) Plasma free fatty acid rate of appearance is increased in men following endurance training. *J. Appl. Physiol.* **86**, 2097–2105

14. Connett, R.J., Honig, C.R., Gayeski, T.E.J. and Brooks, G.A. (1990) Defining hypoxia: a systems view of V_{O_2}, glycolysis, energetics and intracellular PO_2. *J. Appl. Physiol.* **68**, 833–842

15. Randle, P.J., Newsholme, E.A. and Garland, P.B. (1964) Effects of fatty acids, ketone bodies and pyruvate and of alloxan-diabetes and starvation on the uptake and metabolic fate of glucose in rat heart and diaphragm muscles. *Biochem. J.* **93**, 652–665

16. Saddik, M., Gamble, J., Witters, L.A. and Lopaschuk, G.D. (1993) Acetyl-CoA carboxylase regulation of fatty acid oxidation in the heart. *J. Biol. Chem.* **268**, 25836–25845

17. Roberts, A.C., Reeves, J.T., Butterfield, G.E., Mazzeo, R.S., Sutton, J.R., Wolfel, E.E. and Brooks, G.A. (1996) Altitude and β-blockade augment glucose utilization during exercise. *J. Appl. Physiol.* **80**, 605–615

18. Turcotte, L.P., Rovner, A.S., Roark R.R. and Brooks, G.A. (1990) Glucose kinetics in gluconeogenesis-inhibited rats during rest and exercise. *Am J. Physiol.* **258**, E203–E211

19. Issekutz, Jr, B. and Miller, H.I. (1962) Plasma free fatty acids during exercise and the effect of lactic acid. *Proc. Soc. Exp. Biol. Med.* **110**, 237–245

20. Barnard, R., Edgerton, V.R. and Peter, J.B. (1970) Effect of exercise on skeletal muscle I. Biochemical and histochemical properties I. *J. Appl. Physiol.* **28**, 762–766

21. Brooks, G.A. (1985) Lactate: Glycolytic end product and oxidative substrate during sustained exercise in mammals — The 'Lactate Shuttle'. In *Circulation, Respiration and Metabolism: Current Comparative Approaches* (Gilles, R., ed.), pp. 208–218, Springer-Verlag, Berlin

22. Brooks, G.A. (1991) Current concepts in lactate exchange. *Med. Sci. Sports Exercise.* **23**, 895–906

23. Kirkwood, S.P., Packer, L. and Brooks, G.A. (1987) Effects of endurance training on a mitochondrial reticulum in limb skeletal muscle. *Arch. Biochem. Biophys.* **255**, 80–88

24. Karlsson, J. and Saltin, B. (1971) Muscle glycogen utilization during work of different intensities. In *Muscle Metabolism During Exercise* (Pernow, B. and Saltin, B., eds.), pp. 289–299, Plenum Press, New York

25. Donovan, C.M. and Brooks, G.A. (1983) Endurance training affects lactate clearance, not lactate production. *Am. J. Physiol.* **244**, E83–E92

26. Deuster, P.A., Chrousos, G.P., Luger, A., DeBolt, J.E., Bernier, L.L., Trostmen, V.H., Kyle, S.B., Montgomery, L.C. and Loriaux, D.L. (1989) Hormonal and metabolic responses of untrained, moderately trained, and highly trained men to three exercise intensities. *Metabolism* **38**, 141–148

27. Davies, K.J.A., Packer, L. and Brooks, G.A. (1981) Biochemical adaptation of mitochondria, muscle and whole-animal respiration to endurance training. *Arch. Biochem. Biophys.* **209**, 539–554

28. Holloszy, J.O. (1967) Biochemical adaptations in muscle. *J. Biol. Chem.* **242**, 2278–2282

29. Molé, P.A., Oscai, L.B. and Holloszy, J.O. (1971) Adaptation of muscle to exercise: increase in levels of palmityl CoA-synthase, carnitine palmityl transferase and palmityl CoA dehydrogenase and the capacity to oxidize fatty acids. *J. Clin. Invest.* **50**, 2323–2330

30. Dudley, G.A., Tullson, P.C. and Terjung, R.C. (1987) Influence of mitochondrial content on the sensitivity of respiratory control. *J. Biol. Chem.* **262**, 9109–9114

30a. Azevedo, J.L., Lehman, S.L., Linderman, J.K. and Brooks, G.A. (1998) Training decreases glycogen turnover during exercise. *Eur. J. Appl. Physiol.*, **78**, 479–486

31. Donovan, C.M. and Sumida, K.D. (1990) Training improves glucose homeostasis in rats during exercise via glucose production. *Am. J. Physiol.* **258**, R770–R776

32. Sumida, K.D. and Donovan, C.M. (1993) Enhanced gluconeogenesis from lactate in perfused livers after endurance training. *J. Appl. Physiol.* **74**, 782–787

33. Green, H.J., Helyar, R., Ball-Burnett, M., Kowalchuk, N., Symon, S. and Farrance, B. (1992) Metabolic adaptations to training precede changes in muscle mitochondrial capacity. *J. Appl. Physiol.* **72**, 484–4191

34. Clutter, W.E., Bier, D.M., Shah, S.D. and Cryer, P.E. (1980) Epinephrine plasma metabolic clearance rates and physiologic thresholds for metabolic and hemodynamic actions in man. *J. Clin. Invest.* **66**, 94–101

35. Sutton, J. and Lazarus, L. (1976) Growth hormone in exercise: comparison of physiological and pharmacological stimuli. *J. Appl. Physiol.* **41**, 523–527

36. Lehmann, M., Wybitul, K., Spori, U. and Keul, J. (1982) Catecholamines, cardiocirculatory, and metabolic response during graduated and continuously increasing exercise. *Int. Arch. Occup. Environ. Health.* **50**, 261–271

37. Gould, A.G. and Dye, J.A. (1932) *Exercise and its Physiology*, p. 91, A.S. Barnes and Co, New York

38. Hill, A.V. and Lupton, H. (1923) Muscular exercise, lactic acid and the supply and utilization of oxygen. *Q. J. Med.* **16**, 135–171

39. Foster, D.W. (1984) From glycogen to ketones — and back. *Diabetes* **33**, 1188–1199

40. Mazzeo, R.S., Brooks, G.A., Schoeller, D.A. and Budinger, T.F. (1986) Disposal of [1-^{13}C]lactate during rest and exercise. *J. Appl. Physiol.* **60**, 232–241

41. Stanley, W.C., Wisneski, J.A., Gertz, E.W., Neese, R.A. and Brooks, G.A. (1988) Glucose and lactate interrelations during moderate intensity exercise in man. *Metabolism* **37**, 850–858

42. Stanley, W.C., Gertz, E.W., Wisneski, J.A., Morris, D.L., Neese, R. and Brooks, G.A. (1986) Lactate metabolism in exercising human skeletal muscle: Evidence for lactate

extraction during net lactate release. *J. Appl. Physiol.* **60**, 1116–1120

43. Roth, D.A. and Brooks, G.A. (1990) Lactate and pyruvate transport is dominated using a pH gradient-sensitive carrier in rat skeletal muscle sarcolemmal vesicles. *Arch. Biochem. Biophys.* **279**, 386–394

44. Gertz, E.W., Wisneski, J.A., Neese, R., Bristow, J.A., Searle, G.L. and Hanlon, J.T. (1981) Myocardial lactate metabolism: evidence of lactate release during net chemical extraction in man. *Circulation* **63**, 1273–1279

6

Water and electrolyte loss and replacement in the exercising human

R.J. Maughan

University Medical School, Foresterhill, Aberdeen AB25 2ZD, U.K.

Introduction

Body temperature and hydration status must be maintained within narrow limits, and there are few situations that challenge the body's homeostatic mechanisms as severely as the performance of hard exercise in a hot climate. Thermoregulation and the control of water and electrolyte balance are intimately linked in this situation, and both may fail with catastrophic results if the challenge is too severe. The problem is often seen at its most acute in sport, where the demands of competition mean that the individual athlete — who is highly trained and, therefore, capable of sustaining a high power output for prolonged periods of time — works to the extreme of his or her physiological capacity, and where the environmental conditions may be unfavourable. Military exercises, where hard work may have to be sustained for prolonged periods in desert conditions with limited water availability, can also present an extreme challenge.

Mild levels of dehydration will impair exercise capacity and prevent the athlete from achieving optimum performance; although water losses are barely sufficient to stimulate the subjective sensation of thirst, exercise performance is likely to be impaired. Even when every effort is made to ensure full hydration, athletes in events lasting more than a few minutes usually perform less well in the heat than in cooler conditions: it is easy to demonstrate that, for endurance events at major athletics championships, times are generally poor when ambient temperature and humidity are high. Performance in short-duration events may also be adversely affected. Severe dehydration and hyperthermia are potentially fatal.

Exercise in the dehydrated state leads to a rapid elevation of body temperature and the onset of heat illness. Where hard exercise has to be performed, and especially in hot conditions, there is consequently a real need to ensure adequate fluid intake before, during and after exercise. This immediately raises questions as to what constitutes an adequate intake, and the type of fluid that should be consumed. The answers are not simple, for the fluid requirement will depend on the intensity and duration of the exercise, the ambient climatic conditions and also on the individual physiology and biochemistry of the individual. Prescriptions for fluid replacement require that the factors which govern the body's temperature and fluid homeostasis, as well as the factors which influence fluid replacement, be understood.

Exercise in the heat

Exercise performance, at least in events lasting more than a few minutes, is poorer at high ambient temperatures than at lower temperatures. In cycle-ergometer exercise at a power output requiring 70% of maximum oxygen uptake ($\dot{V}_{O_2max}$), it has been shown that subjects who were able to exercise for about 92 min at an ambient temperature of 11°C could only continue to exercise for 82 min when the ambient temperature was increased to 21°C: a further increase in the ambient temperature to 30°C caused the subjects to become exhausted after only 52 min of exercise [1a]. Many other studies have produced comparable results.

At the same power output, exercise in the heat results in a higher heart rate and a higher cardiac output, as well as higher core and skin temperatures, compared with exercise in a cooler environment. There are also some metabolic differences: exercise in the heat is usually accompanied by a higher blood lactate concentration and there is some evidence of a faster rate of depletion of muscle glycogen. These cardiovascular and metabolic alterations are accompanied by a greater subjective sensation of effort in the heat. The response to exercise in the heat is, therefore, determined in part by the power output and in part by the degree of heat stress.

Some measure of heat stress is required to quantify the effects of environmental conditions on physiological function, and a number of different heat-stress indices are used. The heat stress depends on both ambient temperature and humidity: at high temperatures where the body relies on evaporative heat loss to prevent hyperthermia, humidity becomes increasingly important. When the humidity is high, evaporation is prevented or at least reduced, and the ability to limit development of hyperthermia is lost.

Perhaps the most useful index of heat stress is given by the wet bulb globe temperature (WBGT), which incorporates the wet bulb temperature (T_{wb}), which incorporates a humidity factor; the black globe temperature (T_g), which takes account of the radiant energy; and the dry bulb temperature (T_{db}). From the weighting of these factors in the calculation of the WBGT, it is apparent that a greater loading in this heat-stress index is given to the humidity component:

$$WBGT = (0.7\ T_{wb}) + (0.2\ T_g) + (0.1\ T_{db})$$

Fluid loss during exercise

Fluid loss during exercise is linked to the need to maintain body temperature within narrow limits. Body temperature must be maintained within only a few degrees of the normal resting value of about 37 °C, and the resting individual is normally able to achieve this in a wide range of environmental conditions by a combination

of behavioural and physiological responses. At rest, the rate of energy turnover is low and heat production is correspondingly low; the resting oxygen consumption for an individual with a body mass of about 60–70 kg is around 250 ml/min, corresponding to a rate of heat production of about 70 W. This heat is a by-product of the chemical reactions involved in the conversion of food and stored fuel to usable energy and of all the other metabolic reactions occurring in the body: less than 25% of the energy liberated in the degradation of food-stuffs is available to do useful work. In the resting individual, most of the metabolic work consists of movement of ions across membranes to maintain chemical and ionic gradients.

The skin temperature is normally lower than that of the deep tissues, allowing a gradient from the body core to the periphery: this means that warm blood moving from the heart to the skin will carry heat by convection to the body surface where it can be lost. When the skin temperature is higher than the environmental temperature, heat will be lost to the environment by physical transfer, involving conduction, convection and radiation (Fig. 1). For this to occur, the skin temperature must be greater than the environmental temperature: this is normally true, but, for the unclothed individual, the skin temperature is very much influenced by environmental temperature. This has the effect of reducing the gradient for heat loss from the skin, although at low temperatures it increases the temperature gradient from the core to the body surface: core temperature is little affected by environmental temperature, but skin temperature can fluctuate widely. On a hot day, the gradient between skin and environment may be reversed. When the ambient temperature is higher than skin temperature, heat will be gained from the environment by physical transfer, and the core temperature will begin to rise. The effectiveness of these physical heat-transfer processes is perhaps best demonstrated by the rapid changes that occur in body core temperature in response to immersion in hot or cold water.

During exercise, the rate of heat production can be increased to many times the resting level, and in a simple activity, such as running, the rate of heat production is directly related to the running speed. The total amount of work done depends on the distance covered, the body mass and the mechanical efficiency, but the power output — the rate of work — is critically dependent on speed and mass. Running a marathon in 2 h 30 min requires an average oxygen consumption of about 4 litre/min to be sustained throughout the race for a runner with a body mass of 70 kg [1]. The rate of heat production is, therefore, about 16 times greater than at rest, and the rate of heat loss must be increased equally if body temperature is to be prevented from rising. The total energy cost of running a marathon depends very much on body mass, and amounts to about 1 kcal/kg per km (1 J, the SI unit of work or energy, is approximately equal to 0.25 'Calories'). In races over shorter distances than the marathon, the speed is faster and the rate of heat production is correspondingly higher. A survey of the litera-ture on heat illness in athletic events reveals that hyperthermia may be relatively more common among participants in races over distances of 10–20 km than in longer races.

In spite of the high rates of heat production that must be sustained, marathon runners are normally able to complete a race and yet maintain body temperature within 2–3 °C of the resting level, indicating that heat is being lost from the body almost as fast as it is being produced. Only rarely does this thermoregulatory function fail, but when it does so the results are potentially disastrous. The spectacular collapses which are occasionally observed at major marathon competitions are rather rare when the weather is cool, but become more frequent when it is hot and humid, and are generally the result of a failure of the thermoregulatory system.

At high ambient temperatures, heat is gained by physical transfer from the environment, and the only mechanism by which heat

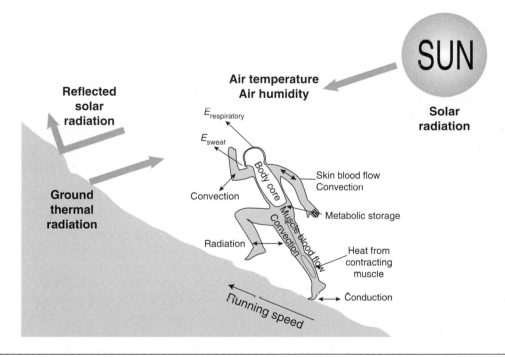

Fig. 1. During intense exercise in the heat, metabolic heat production and heat gain by physical transfer from the environment tend to cause body temperature to rise

Heat loss occurs by evaporation of sweat from the skin (E_{sweat}) and, to a limited extent, of water from the respiratory tract ($E_{respiratory}$).

can be lost from the body is evaporation. Because of the high latent heat of vaporization of water, evaporation of 1 litre of water from the skin will remove 2.4 MJ (580 kcal) of heat from the body. For the 2 h 30 min marathon runner, with a body mass of 70 kg, to balance the rate of heat production by evaporative loss alone would, therefore, require sweat to be evaporated from the skin at a rate of about 1.6 litre/h: at such high sweat rates, an appreciable fraction drips from the skin without evaporating, and a sweat secretion rate of about 2 litre/h is likely to be necessary to achieve this rate of evaporative heat loss. Such high sweat rates are possible, but would result in the loss of 5 litres of body water over the course of the race, corresponding to a loss of more than 7% of body mass for a 70 kg runner. Water that drips from the skin surface without evaporation does not promote heat loss, but will exacerbate the dehydration.

Water will also be lost by evaporation from the respiratory tract, although this is not such an important thermoregulatory mechanism in humans as it is in some animals. Animals with hairy or woolly skin, such as the dog or sheep, clearly have a limited capacity for the evaporation of sweat as a local microclimate develops close to the skin with humidity close to 100%. During hard exercise in a hot dry environment, significant amounts of fluid will be lost from the respiratory tract, as the expired air is saturated with water vapour even at high respiratory rates, but the total loss is small compared with sweat loss. The rise of 2–3°C in body temperature which normally occurs during marathon running means that some of the heat produced is stored, but the effect on heat balance is minimal: a rise in mean body temperature of 3°C for a 70 kg runner would reduce the total requirement for evaporation of sweat by less than 300 ml.

▶ Exercise performance is poorer at higher ambient temperatures due to a combination of cardiovascular and metabolic changes: higher heart rate; increased cardiac output; blood lactate concentration; and an increase in the rate of depletion of muscle glycogen.

▶ Heat stress is dependent on the ambient temperature and humidity and can be determined using the WBGT.

▶ During exercise, the rate of heat production can be increased to many times the resting level. The rate of heat loss must be increased to compensate to prevent body temperature from rising.

▶ At high ambient temperatures, heat is lost from the body by evaporation of sweat; evaporation of 1 litre of water from the skin removes 2.4 MJ (580 kcal) of heat from the body. At high sweat rates, sweat drips from the skin surface without evaporating — this does not promote heat loss but instead exacerbates dehydration.

▶ Many animals lose significant amounts of fluid from the respiratory tract during expiration, but the amount lost by this route is relatively small in humans.

Control of sweating

Sweating is initiated in response to increases in skin and core temperature. The hypothalamus effectively acts as a thermostat and initiates thermoregulatory responses to changes in body temperature. Temperature receptors located in the skin are well placed to detect changes in ambient temperature, and temperature-sensitive neurons in the spinal cord and in the hypothalamus can detect changes in core temperature. If the core temperature increases above the set point, the hypothalamus initiates heat-loss mechanisms. The sweat glands are stimulated to increase sweat production and thus increase the rate of evaporative heat loss. Evaporation of this sweat is promoted by increasing skin blood flow: this is achieved by a reduction in the vasoconstrictor tone in the skin vasculature. The feedback loop is closed when core temperature begins to fall, reducing the input from the heat-sensitive neurons to the hypothalamus. During hard exercise in the heat, this situation does not usually arise, and a steady state may not be established, leading to a progressive rise in core temperature.

The control mechanisms involved in the regulation of sweat secretion are indicated in Fig. 2.

Effects of sweat loss on exercise performance

The effects of dehydration on endurance capacity have been extensively studied by the military and by sports scientists as well as by those interested in the underlying physiology, and a decrease in exercise performance after dehydration is well documented. The majority of more recent studies, especially those in which exercise was carried out in the heat, have confirmed this finding, and have also shown an increase in serum osmolality, heart rate and core temperature during exercise in dehydrated compared with euhydrated subjects.

It is often reported that exercise performance is impaired when an individual is dehydrated by as little as 2% of body mass, and that losses in excess of 5% of body mass can decrease the capacity for work by about 30%. Athletes in short-duration events often feel that dehydration and the need for rehydration are not problems that need concern them: where the duration of an event is no more than a few minutes, the amount of sweat lost during competition is inevitably small even though the rate

of heat production is very high. However, where the competition is taking place in a hot climate, there is a real danger that the athlete will be dehydrated before the event begins, and performance will suffer if this is the case. The capacity to perform high-intensity exercise which results in exhaustion within only a few minutes has been shown to be reduced by as much as 45% by prior prolonged exercise that resulted in a loss of water corresponding to only 2.5% of the initial body mass: smaller, but substantial, reductions in performance occurred after administration of diuretics or after sweat loss in a sauna [2]. Armstrong et al. [3] showed that, when comparisons were made between the normally hydrated state and after a diuretic-induced dehydration of 2% of body mass, dehydration resulted in performance in simulated races over distances of 1500–10000 m being reduced by an average of about 3–7%: at 1500 m, a deficit of 3% in performance is worth more than 6 s at world-class level!

In more prolonged exercise, performance is impaired in the heat, even when subjects are fully hydrated at the beginning of exercise. This suggests a clear role for some factor related to temperature regulation or fluid balance in the fatigue process, at least when the ambient temperature is high. The effects of variations in the ambient temperature on the time for which a fixed power output could be continued demonstrates the importance of thermoregulation as a possible limiting factor in exercise performance

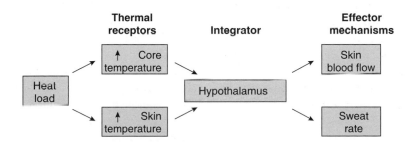

Fig. 2. Regulatory steps in sweat secretion
This is a good example of a simple feedback loop: increasing temperature is detected by temperature receptors that signal the hypothalamus to initiate heat-loss mechanisms. As these take effect, temperature falls and the signal to the hypothalamus decreases.

[1a]. At a power output of about 70% of maximum oxygen uptake, mean exercise time was highest (92 min) at a temperature of 11°C, and was reduced to 53 min when the temperature was increased to 30°C. It is not clear whether the factors causing the subjects to become fatigued and to stop exercising are the same in these different conditions, but when the only variable is the environmental temperature it is clear that this has a profound effect on exercise capacity.

The fluid lost in sweat is derived in varying proportions from the plasma, the extracellular water and the intracellular water space: the distribution of the losses among these spaces depends on the extent of the losses and also on the time-scale over which the loss occurs. In the first few minutes of exercise, there is a sharp fall in the plasma volume. This is due not to water loss from the body, but rather to a redistribution between the water compartments. When exercise begins, the osmolality of the intracellular space in the exercising muscles increases as a result of the increased concentration of glycolytic intermediates and products, especially lactate. In hard exercise, at 70% or more of $\dot{V}_{O_2max}$, the plasma volume may fall by 10–15% within the first 5–10 min of exercise; however, even in the absence of fluid intake, it may begin to return towards the pre-exercise level as exercise proceeds and the distribution of water is again altered. Sweat losses in the first few minutes of exercise are generally negligible, so it is clear that the fall in plasma volume is not caused by water loss from the body. If the exercise is prolonged and the sweat losses are large, the plasma volume may continue to fall, although this can be influenced by the amount and type of fluids consumed.

The decrease in plasma volume that accompanies dehydration may be of particular importance in influencing work capacity; blood flow to the muscles must be maintained at a high level to supply oxygen and substrates, but a high blood flow to the skin is also necessary to convect heat to the body surface where it can be dissipated. When the ambient temperature is high, and blood volume has been decreased by sweat loss during prolonged exercise, there may be difficulty in meeting the requirement for a high blood flow to both these tissues. In this situation, peripheral vasoconstriction will occur to reduce the skin blood flow. This will allow central venous pressure and muscle blood flow to be maintained, but the reduced flow of warm blood to the skin will reduce heat loss and cause body temperature to rise. Maintenance of blood pressure appears to be a priority in the hierarchy of control mechanisms, and can only be achieved by peripheral vasoconstriction when the blood volume is reduced.

Electrolyte losses with sweating

Sweat is often described as an ultrafiltrate of plasma, but the ionic composition of sweat is not only different from that of plasma, it also varies considerably between individuals and over time in any one individual [4]. There are considerable difficulties in establishing the ionic composition of human sweat, as it varies with the sweating rate, with time, with hydration status and with the degree of acclimation. There are also differences in the composition of sweat secreted at different sites on the body, although it is not clear to what extent this is simply a reflection of regional differences in sweating rate. In response to a standard heat stress, the sweat rate increases with training and acclimation and the electrolyte content decreases. These adaptations allow improved thermoregulation while conserving electrolytes. Some literature values for human sweat composition are shown in Table 1.

The major electrolytes in sweat, as in the extracellular fluid, are sodium and chloride, although the sweat concentrations of these ions are invariably lower than those in plasma. Sweat sodium concentration is normally in the range 20–80 mM. Although there is a large inter-individual variation, the composition appears to be relatively consistent for any individual at any given flow rate. The potassium content of sweat (4–8 mM) is high relative to that of plasma (about 4–5 mM) but is low relative to that of the intracellular water (about 150 mM). The concentration of most plasma

Table 1. Normal electrolyte composition of the extracellular space (plasma) and of the intracellular space

	Electrolyte composition (mmol/l)		
	Plasma	Intracellular space	Sweat
Sodium	130–155	10	20–80
Potassium	3.2–5.5	150	4–8
Calcium	2.1–2.9	0	0–1
Magnesium	0.7–1.5	15	<0.2
Chloride	96–110	8	20–60
Bicarbonate	23–28	10	0–35
Phosphate	0.7–1.6	65	0.1–0.2
Sulphate	0.3–0.9	10	0.1–2.0

A range of values for sweat composition is also included, but there may be values which fall outwith this range in some situations. All concentrations are expressed in mmol per litre. The variability in the composition of the intracellular space is less-well understood than is the case for plasma, mainly because of the difficulties involved in obtaining samples of the intracellular fluid. The value for intracellular calcium does not include calcium sequestered within the cell.

electrolytes increases during prolonged exercise that results in sweat loss. Although the concentration of potassium and magnesium in sweat is high relative to that in the plasma, the plasma content of these ions represents only a small fraction of the whole-body stores; Costill and Miller [5] estimated that only about 1% of the body stores of these electrolytes was lost when individuals were dehydrated by 5–8% of body mass.

▶ The hypothalamus acts as a thermostat and initiates thermoregulatory responses to change in body temperature. Sweat glands are stimulated and evaporation is promoted by increasing skin blood flow.

▶ Fluid loss of 2% of body mass impairs performance; losses over 5% of body mass can impair work capacity by ~30%.

▶ Fluid lost during sweating is derived from plasma, extracellular water and intracellular water in proportions that vary according to extent of loss and the time-scale over which it occurs.

▶ At high temperatures, when blood volume has decreased by sweat loss during prolonged exercise, peripheral vasoconstriction reduces skin blood flow, allowing central venous pressure and muscle blood flow to be maintained; however, heat loss will be reduced and body temperature will rise.

▶ The ionic composition of sweat varies between individuals and according to rate of sweating, time, hydration status, degree of acclimation and site of secretion in any individual.

▶ Sweat rate increases with training and acclimation.

▶ The main electrolytes in sweat are sodium and chloride, with smaller amounts of potassium and magnesium.

Control of water and electrolyte balance

The excretion of some of the waste products of metabolism, particularly the nitrogen derived from amino acid degradation, and the regulation of the body's water and electrolyte balance are the primary functions of the kidneys. Excess water or solute is excreted, and where there is a deficiency of water or electrolytes, these are conserved until the balance is restored. Under normal conditions, the osmolality of the extracellular fluid is maintained within narrow limits; since this is strongly influenced by the sodium concentration, sodium and water balance are closely linked. Plasma osmolality is normally about 285–290 mosmol/l, and the plasma sodium therefore accounts for about half of the total osmolality.

At rest, the kidneys receive about 25% of the total cardiac output of about 5 litre/min: approximately 15–20% of the renal plasma flow is continuously filtered out by the glomeruli, resulting in the production of about 170 litres of filtrate per day for a 70 kg individual. Most of this (99% or more) is reabsorbed in the tubular system, leaving about 1–1.5 litres to appear as urine. The volume of urine produced is determined primarily by the action of antidiuretic hormone (ADH). In the absence of ADH, almost all of the water that reaches the distal tubule of the nephron is excreted as urine, and urine output can reach 1.5 litre/h after ingestion of large volumes of electrolyte-free solutions. ADH regulates water reabsorption by increasing the permeability of the distal tubule of the nephron and the collecting duct to water. ADH is released from the posterior lobe of the pituitary in response to signals from the supraoptic nucleus of the hypothalamus: the main stimuli for release of ADH, which is normally present only in low concentrations, are an increased signal from the osmoreceptors located within the hypothalamus, a decrease in blood volume, which is detected by low-pressure receptors in the atria, and by high-pressure baroreceptors in the aortic arch and carotid sinus. An increased plasma angiotensin concentration will also stimulate ADH output.

The sodium concentration of the plasma is regulated by the renal reabsorption of sodium from the glomerular filtrate: this allows regulation which is largely independent of the dietary intake. Most of the reabsorption occurs in the proximal tubule, but active absorption also occurs in the distal tubules and collecting ducts. A number of factors influence the extent to which reabsorption occurs, and among these is the action of aldosterone, which promotes sodium reabsorption in the distal tubules and enhances the excretion of potassium and hydrogen ions. Aldosterone is released from the kidney in response to a fall in the circulating sodium concentration or a rise in plasma potassium: aldosterone release is also stimulated by angiotensin, which is produced by the renin–angiotensin system in response to a decrease in the plasma sodium concentration. Angiotensin thus has a two-fold action, on the release of aldosterone as well as ADH. Atrial natriuretic factor (ANF) is a peptide synthesized in and released from the atria of the heart in response to atrial distension, and a major role for this peptide has been proposed. It increases the glomerular filtration rate and decreases sodium and water reabsorption leading to an increased loss: this may be important in the regulation of extracellular volume, but it seems unlikely that ANF plays a significant role during exercise. Regulation of the body's sodium balance has profound implications for fluid balance, as sodium salts account for more than 90% of the osmotic pressure of the extracellular fluid.

Loss of hypotonic fluid as sweat, which usually has an osmolality of less than about 150 mosmol/kg, during prolonged exercise usually results in a fall in blood and extracellular fluid volume and an increased plasma osmolality: both these changes act as stimuli for the release of ADH [6]. The plasma ADH concentration during exercise has been reported to increase as a function of the exercise intensity. Renal blood flow is also reduced in proportion to the exercise intensity and may be as low as 25% of the resting level during strenuous exercise. These factors combine to result in a decreased urine flow during exercise, and usually for some time afterwards as well. It has been pointed out, however, that the volume of water conserved by this decreased urine flow during exercise is small, probably amounting to no more than 12–45 ml/h [7].

The effect of exercise is normally to decrease the renal excretion of sodium and to increase the excretion of potassium, although the effect on potassium excretion is rather variable [7]. These effects appear to be largely due to an increased rate of aldosterone production during exercise. Although the concentrations of sodium and more especially of potassium in the urine are generally high relative to the concentrations in extracellular fluid, the extent of total urinary losses in most exercise situations is small because of the reduced renal blood flow and the low volumes of urine formed.

The primary drive to replacement of fluid losses and the maintenance of water balance is the subjective sensation of thirst. This is a complex mechanism, responding to changes in the osmolality and the sodium concentration of the plasma, but it is not at present clear that the volume of fluid consumed is closely related to either of these two variables. These two factors are themselves normally closely linked, as sodium is the major cation of the extracellular space (Table 1). Indeed, it has been suggested that the osmoreceptor is a sodium receptor. There is some evidence that voluntary fluid intake in active individuals exposed to a warm environment is not closely linked with any of a wide range of plasma variables, including osmolality and sodium concentration, but is more closely related to the water deficit incurred by urine and sweat losses [8]. One fact that is clear is that the stimulation of thirst in response to exercise and hot, humid conditions is not sufficient to ensure an adequate fluid intake and some degree of hypohydration will ensue unless a conscious effort is made to increase intake.

> ► Under normal conditions, plasma osmolality is conserved within narrow limits, and is largely determined by sodium concentration.
> ► The volume of urine an individual produces is governed by the action of ADH, which is released from the pituitary lobe in response to decreased blood volume and increased plasma angiotensin concentration.
> ► Plasma sodium concentration is regulated by renal absorption of sodium, which is influenced by the action of aldosterone and ANF.
> ► Sweating during prolonged exercise results in a fall in blood and extracellular fluid volume and an increase in plasma osmolality — this stimulates aldosterone release.
> ► The combination of increased plasma ADH concentration and reduced renal blood flow during exercise leads to decreased urine flow during and after exercise.
> ► Exercise usually decreases renal excretion of sodium and increases potassium excretion due to increased aldosterone production.
> ► Thirst is a complex, subjective sensation which may be due to changes in osmolality and sodium levels in plasma and to the water deficit caused by sweating and urine excretion.

Fluid replacement during exercise

The main factors causing fatigue in prolonged exercise are depletion of the carbohydrate stores and dehydration. Therefore, the ability to sustain a high rate of work output requires that an adequate supply of carbohydrate substrate be available to the working muscles, and fluid ingestion during exercise has the twin aims of providing a source of carbohydrate fuel to supplement the body's limited stores and of supplying water to replace the losses incurred by sweating. The rates at which substrate and water can be supplied during exercise are limited by the capacity of the gastro-intestinal tract to deliver these. The rates of gastric emptying and intestinal absorption are both possible limitations, but it is not clear which of these processes is limiting. It is commonly assumed that the rate of gastric emptying will determine the maximum rates of fluid and substrate availability [9,10] but the evidence is not strong. In most cases, fluid replacement is limited not by physiological function but by the volume consumed. Some understanding of gastro-intestinal function is necessary to understand the limitations to water and nutrient supply and to determine the optimum composition of solutions to be ingested in different situations.

There is good evidence that carbohydrate feeding and maintenance of hydration status separately improve exercise performance, and that the effects are additive [11]. The greatest benefit in most situations should, therefore, come from ingestion of carbohydrate-containing drinks, but the optimum balance has been the subject of much debate, and there is still no agreement.

Increasing the carbohydrate content of drinks will increase the amount of fuel which can be supplied, but will tend to decrease the rate at which water can be made available. Even dilute glucose solutions (40 g/l or more) will slow the rate of gastric emptying [12], but active absorption of glucose, which is co-transported with sodium in the small intestine, stimulates water absorption: the highest rates of oral water replacement are thus achieved with dilute solutions of glucose and sodium salts [13]. The highest rates of glucose emptying from the stomach are observed with concentrated glucose solutions (200 g/l or more), even though the volume emptied is small when such solutions are consumed (Fig. 3). The pattern of emptying is exponential rather than linear: the volume of fluid in the stomach is a major factor stimulating emptying, and the rate decreases progressively as the volume falls. In the case of nutrient-containing solutions, feedback from receptors in the small intestine also acts to slow emptying and thus regulate the delivery of nutrients to the sites of absorption.

There is some evidence that the rate of glucose absorption, which takes place in the upper part of the small intestine, is already maximal at moderate glucose concentrations [14]. Glucose is actively absorbed by the cells of the intestinal wall: this is an energy-consuming process, and sodium is co-transported with the glucose molecule. There is no active transport mechanism for water in the gastro-intestinal tract: it simply follows osmotic gradients, and is free to move in either direction across the luminal wall. Solutions containing high concentrations of glucose or any other solute will, because of their high osmolality, stimulate high rates of secretion of water into the gastro-intestinal tract; this will have the effect of accentuating dehydration and may also result in a subjective sensation of discomfort. In extreme cases, the solute load may exceed the absorptive capacity

of the lower bowel, resulting in diarrhoea: this may be particularly likely when fructose is consumed in large amounts, as the absorptive capacity of the intestine for fructose is low compared with that for glucose. Gastro-intestinal distress appears to be relatively common among endurance athletes, and clearly indicates an accelerated rate of gastro-intestinal transit or a reduced absorptive capacity in those susceptible individuals [15].

Where provision of water is the first priority, therefore, the carbohydrate content of drinks will be low, perhaps about 30–50 g/l, even though this may restrict the rate at which substrate is provided. The substrate concentration is limited by the need to maintain the hypotonicity of the luminal contents: it has been argued that this can be achieved by the substitution of disaccharides or higher polymers for glucose, allowing greater carbohydrate provision. Although the evidence is not entirely convincing, there may well be benefits from the use of glucose polymers. The addition of sodium to drinks has been questioned on the grounds that secretion of sodium into the intestinal lumen will occur sufficiently rapidly to stimulate maximal rates of glucose–sodium co-transport. The evidence here is also questionable, but there are other good reasons for addition of sodium to rehydration fluids, as discussed below. The suggestion that gastric emptying of drinks is enhanced if they are chilled before ingestion is not supported by more recent evidence, although the palatability of many drinks, as well as the subject sensation of refreshment, is improved when they are consumed at low temperatures. Given that voluntary intake generally limits fluid replacement, the palatability issue may be crucial in stimulating an increased fluid consumption.

The composition of drinks to be taken will thus be influenced by the relative importance of the need to supply fuel and water; this depends primarily on the ambient conditions and the exercise intensity, but the inter-individual variability in sweating characteristics, and hence in the level of dehydration that will be incurred in any situation, is extremely large. Carbohydrate depletion will result in fatigue and a reduction

(a)

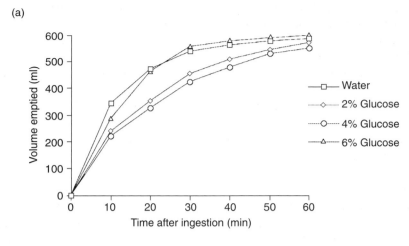

(b)

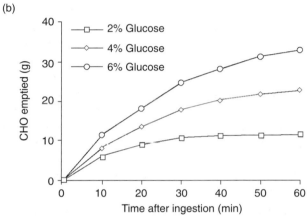

Fig. 3. Gastric emptying of glucose solutions

These two figures show the gastric emptying pattern after ingestion of 600 ml of a test drink containing 0, 2, 4 or 6 g of glucose per 100 ml. Emptying is slowed in proportion to the glucose concentration, but the delivery of glucose is increased as concentration increases in spite of the smaller volume being emptied. (a) Volume of the original drink emptied from the stomach over the 60 min after ingestion. (b) Amount of carbohydrate (CHO) emptied from the stomach over time by the different glucose-containing solutions. Redrawn with permission from [12]. ©1994 Lippincott, Williams & Wilkins.

in the exercise intensity which can be sustained, but is not normally a life-threatening condition. Disturbances in fluid balance and temperature regulation have potentially more serious consequences, and it may be, therefore, that the emphasis for the majority of participants in endurance events should be on proper maintenance of fluid and electrolyte balance.

Effects of fluid ingestion on exercise performance

The effects of feeding different types and amounts of beverages during exercise have been extensively investigated, using a wide variety of experimental models. Laboratory investigations into the ergogenic effects of the administration of such drinks during exercise have usually relied upon changes in physiological function during sub-maximal exercise or on the exercise time to exhaustion at a fixed work rate as a measure of performance. While this is a perfectly valid approach in itself, it must be appreciated that there are difficulties in extrapolating results obtained in this way to a race situation in which the work load is likely to fluctuate as the pace, the weather conditions and the topography vary, and where tactical considerations and motivational factors are involved. It is possible to demonstrate large differences in the time for which a fixed work load can be sustained in laboratory tests when carbohydrate solutions are given during exercise: in one study, for example, a 30% increase (from 3 h to 4 h) was seen. In a simulated race situation, where a fixed distance had to be covered as fast as possible, the advantage would translate to no more than a few percent, and in a real competition it would probably be even less. Even a few percent, however, is often the difference between a world-class performance and a mediocre one, and the difference between first and last in an Olympic final may be much less than this. To take account of some of these factors, some recent investigations have used exercise tests involving intermittent exercise, simulated races or prolonged exercise followed by a sprint finish. Because different exercise tests and different solutions and rates of administration have been used in these various studies, comparisons between studies are difficult. Some studies have included a trial where no fluids were given, whereas others have compared the effects of test solutions with trials where plain water or flavoured placebo drinks were given. These studies have been the subject of a number of extensive reviews which have concentrated on the effects of administration of carbohydrates, electrolytes and water on exercise performance [9,10,14].

Not all of the studies reported in the literature have shown a positive effect of fluid ingestion on performance, but, with the exception of a few investigations where the composition of the drinks administered was such as to result in gastro-intestinal disturbances, there are no studies showing that fluid ingestion will have an adverse effect on performance. It is also true that, in spite of frequently repeated statements that plain water is a suitable beverage for ingestion during exercise, the evidence shows that ingestion of a suitable formulated carbohydrate-electrolyte drink will almost invariably give better results in terms of minimizing the disturbances in fluid and electrolyte balance and in maintaining exercise performance. Some of the reports in the literature show no significant difference between trials where water or carbohydrate-electrolyte drinks are consumed, but the trend is always in favour of the addition of carbohydrate and sodium to drinks. In prolonged exercise, where substrate depletion is likely to occur, or during exercise in the heat which is sufficiently prolonged to result in dehydration, there can be no question that performance is improved by the regular ingestion of suitable glucose-electrolyte drinks.

▶ Fluid ingestion during exercise provides a source of carbohydrate to the working muscles, and supplies water to replace losses as a result of sweating, at a rate determined by the capacity of the gastro-intestinal tract.

▶ Carbohydrate feeding and maintenance of hydration status separately improve exercise performance.

▶ Drink composition will depend on the need to supply fuel or water, which in turn depends on the ambient conditions and exercise intensity, as well as the individual's sweating characteristics. Where water replacement is foremost, the carbohydrate content of drinks should be low.

▶ Evidence shows that ingestion of a suitably formulated carbohydrate-electrolyte drink gives better results that ingestion of pure water — in terms of minimizing disturbance of fluid and electrolyte balance and in maintaining exercise performance.

Restoration of water and electrolyte balance after exercise

Replacement of water and electrolyte losses in the post-exercise period may be of crucial importance for maintenance of exercise capacity when repeated bouts of exercise have to be performed. Restoration of euhydration and re-establishment of normal electrolyte levels are also a normal part of the recovery process when significant sweat losses have occurred. The need for replacement will obviously depend on the extent of the losses incurred during exercise, but the time-scale may be influenced by the time and nature of subsequent exercise bouts. Rapid rehydration may be important in events such as wrestling, boxing and weightlifting where competition is by body-weight category. Competitors in these events frequently undergo acute thermal and exercise-induced dehydration to make weight, and some competitors may lose as much as 10% of body mass in the last day or two before competition. The time interval between the weigh-in and competition is different in different sports, but restoration of losses on such a scale within a short period of time presents a major challenge.

Ingestion of large volumes of plain water normally leads to a fall in the circulating sodium concentration and in the plasma osmolality. It has sometimes been reported that excessive intake of fluids with a low sodium content can induce hyponatraemia — defined as a plasma sodium concentration below the renal range of 130–155 mmol/litre — during exercise of long duration. Ingestion of plain water in the post-exercise period also results in a rapid fall in the plasma sodium concentration and in plasma osmolality [16]. These changes have the effect of reducing the stimulus to drink (thirst) and of increasing the rate of urine output, both of which will delay the rehydration process. In one study, subjects exercised at low intensity in the heat for 90–110 min, inducing a mean dehydration of 2.3% of body mass, and then rested for 1 h before beginning to drink [16]. Plasma volume was not restored until after 60 min when plain water was ingested together with placebo (sucrose) capsules. In contrast, when sodium chloride capsules were ingested with water to give a saline solution with an effective concentration of 0.45% (77 mM), plasma volume was restored within 20 min. In the sodium chloride trial, voluntary fluid intake was higher and urine output was less; 71% of the water loss was retained within 3 h compared with 51% in the plain water trial. The delayed rehydration in the water trial appeared to be a result of a loss of sodium, accompanied by water, in the urine caused by enhanced plasma renin activity and aldosterone levels.

It is clear from the results of these studies that rehydration after exercise can only be achieved if the sodium lost in sweat is replaced as well as the water, and more recent studies have confirmed that the restoration of euhydration after exercise requires replacement of electrolytes (primarily sodium) as well as water losses. If no fluid is consumed after sweating, continuing urine formation will result in further dehydration. If a fixed volume of fluid (in excess of the volume lost) is consumed after exercise-induced dehydration, some will be lost in the urine within the first few hours, and if the urine losses are large, full rehydration will not be achieved. The fraction of the ingested fluid that is lost in urine within the first few hours of the recovery period is inversely related to the sodium content, within the range 0–100 mM sodium [17].

It might be suggested that rehydration drinks should have a sodium concentration similar to that of sweat, but the electrolyte content of sweat varies widely, and no single formulation will meet this requirement for all individuals in all situations. The upper end of the normal range for sodium concentration (80 mM), however, is similar to the sodium concentration of many commercially produced oral rehydration solutions (ORS) intended for use in the treatment of diarrhoea-induced dehydration, and some of these are not unpalatable. The ORS recommended by the World Health Organization for rehydration in cases of severe diarrhoea has a sodium content of 90 mM, reflecting the high sodium losses which may occur in this condition. It also has a high (25 mM) potassium content, again reflecting the high potassium content of the stool lost in tropical diarrhoea. By contrast, the sodium content of most sports drinks is in the range of 10–25 mM and is even lower in some cases; most commonly consumed soft drinks contain virtually no sodium and these drinks are therefore unsuitable when the need for rehydration is crucial. The problem with high sodium concentrations is that this may exert a negative effect on taste, resulting in a reduced consumption. Consumption of an adequate volume remains a priority, and palatable drinks will increase the probability of a sufficient volume being consumed.

Replacement of potassium is also important after exercise: potassium losses in sweat are small, but significant loss may occur in the urine after exercise. The plasma potassium concentration normally increases during exercise. There is a loss of potassium from the active muscle, and some is also released from the liver as its glycogen store falls. Potassium release from cells damaged during exercise, including both muscle and red blood cells, also contributes to the rise in the serum concentration. In most cases, replacement of these losses is achieved through the normal food intake, especially fruit and fruit juices, which are high in potassium, and specific supplementation is not necessary.

▶ Ingesting large volumes of plain water before and after exercise can decrease plasma sodium concentration and plasma osmolality — thus reducing thirst and increasing urine output, both of which delay rehydration.
▶ Rehydration after exercise can only be achieved if electrolytes as well as water loss are replaced.
▶ The 'ideal' rehydration drink would have an electrolyte composition similar to that of sweat; however, due to the variability of sweat composition between individuals, no one formulation would be suitable for all.
▶ Replacement of potassium lost in sweat and urine is important after exercise.

Heat acclimation

Repeated exposure to heat results in a number of physiological and behavioural adaptations that improve an individual's capacity for exercise in these conditions. The process of adaptation to a single environmental stress, such as heat, is generally referred to as acclimation: the term acclimatization is preferred for the response to the many factors that change when an individual moves to a different location with a hot climate.

The magnitude of the adaptation to heat that occurs is closely related to the degree of heat stress to which the individual is exposed. This is obviously determined by ambient conditions, exercise intensity, exercise duration, and other factors such as the amount of clothing worn. Some degree of adaptation is induced by passive heat exposure, such as occurs during sitting in a sauna, but the high temperatures that are involved mean that the exposure time is short. Adaptation seems to be most effectively accomplished by exercise lasting about 60–100 min per day, and will occur even when the remainder of the day is spent in cool conditions. Some adaptation is seen within the first few days and adaptation is more or less complete for most individuals within about 7–14 days. The adaptations that occur are gradually reversed on returning to temperate conditions. It is equally clear that regular endurance training in temperate conditions confers some protection: trained subjects are already partially adapted, but full adaptation is not seen unless a period of time is spent training in the heat.

The main adaptations that occur in response to repeated exercise in the heat are an increased plasma volume, increased sweating rate, an increased ability to sustain high sweating rates, a better distribution of sweating over the whole body surface and a reduced sweat electrolyte content. Sweat sodium concentration normally increases as the sweat rate increases, due to the reduced time available for reabsorption of sodium during passage through the sweat duct, but changes in the sweat gland ultrastructure allow high sweating rates to be achieved without corresponding increases in the electrolyte content.

Among the benefits of endurance training is an expansion of the plasma volume [18]. Although this condition is recognized as a chronic state in the endurance-trained individual, an acute expansion of plasma volume occurs in response to a single bout of strenuous exercise: this effect is apparent within a few hours of completion of exercise and may persist for several days. This post-exercise hypervolaemia should be regarded as an acute response rather than an adaptation, although it may appear to be one of the first responses to occur when an individual embarks on a training regimen. Circulating electrolyte and total protein concentrations are normal in the endurance-trained individual in spite of the enlarged vascular and extracellular spaces, indicating an increased total circulating content.

The increased resting plasma volume in the trained state allows the endurance-trained individual to maintain a higher total blood volume during exercise, allowing for better maintenance of cardiac output, albeit at the cost of a lower circulating haemoglobin concentration. In addition, the increased plasma volume is associated with an increased sweating rate, which limits the rise in body temperature. These adaptive responses appear to occur within a few days of exposure to exercise in the heat, although, as pointed out above, this may not necessarily be a true adaptation. The time-course of these changes in men exposed to exercise (40–50% $\dot{V}_{O_2max}$ for 4 h) in extreme heat (45°C) for 10 days shows that, although there were marked differences between individuals in their responses, the resting plasma volume increased progressively over the first 6 days, reaching a value about 23% greater than the control, with little change thereafter. The main adaptations, in terms of an increased sweating rate and an improved thermoregulatory response (with body temperature lower by 1°C and heart rate lower by 30 beats per min in the later stages of exercise), occurred slightly later than the cardiovascular adaptations, with little change in the first 4 days.

Although there is clear evidence that acclimation by exercise in the heat over a period of several days will improve the thermoregulatory response during exercise, this does not reduce the need to replace fluids during the exercise period. Better maintenance of body temperature is achieved at the expense of an increased water (sweat) loss, and this loss must be replaced. Although the increased sweating rate allows for a greater evaporative heat loss, the proportion of the sweat which is not evaporated and which therefore drips wastefully from the skin is also normally increased. A high sweat rate may be necessary to ensure adequate evaporative heat loss, but it does seem that many individuals have an inefficient sweating mechanism: even in the unacclimated state their rate of sweat secretion appears to greatly exceed the maximum evaporative capacity. Heat acclimation will, therefore, actually increase the requirement for fluid replacement because of the enhanced sweating response. Some studies have shown that heat-adapted individuals will voluntarily increase their fluid intake during heat stress, but it is not clear whether this response reflects an increased sensitivity of the thirst mechanism or whether it can be explained by a learning effect.

If dehydration is allowed to occur, the improved ability to tolerate heat which results from the acclimation process will disappear completely: in other words, regular exposure to exercise in the heat does not result in an increased tolerance to dehydration or to hyperthermia [19]. It has been demonstrated in a number of studies that there is no adaptation to dehydration: there is, therefore, no reason to restrict fluid intake in individuals exposed to heat stress.

▶ Repeated exposure to heat results in physiological and behavioural adaptations that improve the capacity for exercise under these conditions.
▶ Heat adaptation is most effective when exercise is performed for 60–100 min per day over a period of 7–14 days.

▶ The main adaptations that occur are an increased plasma volume, increased sweating rate, increased ability to maintain a high sweating rate, reduced electrolyte content in sweat and better distribution of sweat over the body surface.

▶ Endurance training leads to an expansion of plasma volume, which enables a higher blood volume to be maintained during exercise, which in turn allows increased sweating rate, thus limiting rises in body temperature.

▶ Heat acclimation increases requirements for fluid replacement due to the enhanced sweating response.

Conclusions

▶ Exercise in the heat causes body temperature to rise and sweating to be invoked. This helps thermoregulation but leads to water and electrolyte losses.

▶ Exercise performance is reduced when the ambient temperature and humidity are high.

▶ Exercise performance is impaired by dehydration, and fluid replacement can reduce the impact of adverse environmental conditions on performance.

▶ In prolonged exercise, the blood volume normally falls, and the cardiovascular system is faced with demands for a high muscle blood flow and high skin blood flow.

▶ The major electrolyte lost in sweat is sodium; although the sweat sodium concentration is low relative to that in plasma, it is much higher than the intracellular sodium concentration.

▶ Drinks containing low levels of glucose and sodium salts are most effective in promoting water replacement and also supply substrate for the working muscles.

▶ Post-exercise rehydration requires replacement of sweat electrolyte losses as well as volume replacement.

▶ Exercise performance and thermoregulatory capacity in the heat are improved by prior acclimation.

References

1. Maughan, R.J. and Leiper, J.B. (1983) Aerobic capacity and fractional utilisation of aerobic capacity in elite and non-elite male and female marathon runners. *Eur. J. Appl. Physiol.* 52, 80–87

1a. Galloway, S.D.R. and Maughan, R.J. (1997) Effects of ambient temperature on the capacity to perform prolonged cycle exercise in man. *Med. Sci. Sports Exercise* 29, 1240–1249

2. Nielsen, B., Kubica, R., Bonnesen, A., Rasmussen, I.B., Stoklosa, J. and Wilk B. (1981) Physical work capacity after dehydration and hyperthermia. *Scand. J. Sports Sci.* 3, 2–10

3. Armstrong, L.E., Costill, D.L. and Fink, W.J. (1985) Influence of diuretic-induced dehydration on competitive running performance. *Med. Sci. Sports Exercise* 17, 456–461

4. Costill, D.L. (1977) Sweating: its composition and effects on body fluids. *Ann. NY Acad. Sci.* 301, 160–174

5. Costill, D.L. and Miller, J.M. (1980) Nutrition for endurance sport. *Int. J. Sports Med.* 1, 2–14

6. Castenfors, J. (1977) Renal function during prolonged exercise. *Ann. NY Acad. Sci.* 301, 151–159

7. Zambraski, E.J. (1990) Renal regulation of fluid homeostasis during exercise. In *Perspectives in Exercise Science and Sports Medicine* vol. 3 (Gisolfi, C.V. and Lamb, D.R., eds.), pp. 247–280, Benchmark Press, Carmel

8. Greenleaf, J.E., Averkin, E.G. and Sargent, F. (1966) Water consumption by man in a warm environment: a statistical analysis. *J. Appl. Physiol.* 21, 93–98

9. Lamb, D.R. and Brodowicz, G.R. (1986) Optimal use of fluids of varying formulations to minimize exercise-induced disturbances in homeostasis. *Sports Med.* 3, 247–274

10. Murray, R. (1987) The effects of consuming carbohydrate-electrolyte beverages on gastric emptying and fluid absorption during and following exercise. *Sports Med.* 4, 322–351

11. Below, P., Mora-Rodriguez, R., Gonzalez-Alonso, J. and Coyle, E.F. (1995) Fluid and carbohydrate ingestion independently improve performance during 1 h of intense cycling. *Med. Sci Sports Exercise* 27, 200–210

12. Vist, G.E. and Maughan, R.J. (1994) The effect of increasing glucose concentration on the rate of gastric emptying in man. *Med. Sci Sports Exercise* 26, 1269–1273

13. Maughan, R.J. (1991) Fluid and electrolyte loss and replacement in exercise. *J. Sports Sci.* 9, 117–142

14. Leiper, J.B. and Maughan, R.J. (1988) Experimental models for the investigation of water and solute transport in man: implications for oral rehydration solutions. *Drugs* 36(4), 65–79

15. Brouns, F., Saris, W.H.M. and Rehrer, N.J. (1987) Abdominal complaints and gastrointestinal function during long-lasting exercise. *Int. J. Sports Med.* 8, 175–189

16. Nose, H., Mack, G.W., Shi, X. and Nadel, E.R. (1988) Role of osmolality and plasma volume during rehydration in humans. *J. Appl. Physiol.* 65, 325–331

17. Maughan, R.J. and Leiper, J.B. (1995) Effects of sodium content of ingested fluids on post-exercise rehydration in man. *Eur. J. Appl. Physiol.* 71, 311–319

18. Hallberg, L. and Magnusson, B. (1984) The aetiology of sports anaemia. *Acta Med. Scand.* 216, 145–148

19. Sawka, M.N., Pandolf, K.B. (1990) In *Perspectives in Exercise Science and Sports Medicine* vol. 3 (Gisolfi, C.V. and Lamb, D.A., eds.), pp. 1–38, Benchmark, Carmel

7

Cardiovascular function and oxygen delivery during exercise

Niels H. Secher

The Copenhagen Muscle Research Centre, Department of Anaesthesia, Rigshospitalet 2034, University of Copenhagen, Blegdamsvej 9, DK-2100 Copenhagen Ø, Denmark

Introduction

The increase in metabolism during exercise requires that substrate and oxygen are transported rapidly to working skeletal muscle and the heart. Although cardiovascular adaptation to exercise begins at the onset of exercise, it takes several minutes before the muscles are provided with sufficient blood to balance their demand for oxygen, glucose and fat. In normal humans this is not a problem with respect to substrates for at least 1 h after the onset of exercise because of the large quantity of glycogen that is stored in the muscle. However, before a steady-state of oxygen uptake is established, metabolism is covered by the *oxygen deficit* (the depletion of oxygen stored in haemoglobin and myoglobin) and depletion of muscle phosphagens as well as formation of lactate.

This chapter describes how the circulation delivers the oxygen required to the working muscle, and how exercise influences blood flow to other organs. It also considers whether the circulation limits maximal oxygen uptake ($\dot{V}_{O_2max}$), and how the $\dot{V}_{O_2max}$ is influenced by altitude and physical training.

Cardiovascular regulation during exercise

The circulatory adaptation to exercise is too rapid to be caused by humoral factors. Early investigators (T.A. Aulo, J.E. Johansson and G. Mansfeld) thought it likely that the circulatory (and ventilatory) response to exercise was developed from parallel activation of command signals from the brain to skeletal muscle and the autonomic nervous system (central command), and considered the possibility of input from mechano- or chemoreceptors in the working muscle (the exercise pressor reflex) as well as the effect of temperature.

Central command

The neurophysiological evidence for parallel innervation of the voluntary and the autonomic nervous systems (feed-forward control) is beginning to be established. Evidence is derived from stimulation not only of the hypothalamic 'defence area', but also of the 'field of Forel' and the olive, which increases heart rate and blood pressure. Special interest has been directed to 'the subthalamic locomotor area', which, when stimulated in the paralysed cat, cannot develop reflex activation of the brain stem [1]. In this preparation, stimulation of the subthalamic locomotor area also induces increases in sympathetic nerve activity and hormonal variables associated with exercise. Furthermore, either anaesthesia or destruction of subthalamic areas eliminates the increases in heart rate and blood pressure, and also the adrenaline and corticosterone responses associated with voluntary exercise that give rise to glucose mobilization. The cerebellum should also be considered in relation to the command integration of motor and autonomic nerve activity. This is relevant because neural activity is present in the cerebellum before it can be detected in the cortex, and in support of a role for the cerebellum in cardiovascular regulation, stimulation of the anterior vermis inhibits the sinus reflex. It is, therefore, unlikely that central command simply represents 'cortical irradiation' as proposed by Krogh and Lindhard [2], but cortical areas may be of importance for the early increase in muscle blood flow.

The exercise pressor reflex

In animal experiments, skeletal muscle activation by way of stimulation of the cut ventral roots of the spinal nerves generates increases in heart rate, blood pressure and ventilation (feedback control) [3]. Although such responses are most often demonstrated during isometric contraction, they are also present during rhythmic contractions. The responses are known to be a consequence of the muscle contraction because the preparation allows no retrograde activation of the cardiovascular control areas in the brain stem, and curarization of the animal abolishes the response (Fig. 1). The neural transmission

of the cardiovascular response is confirmed, because it is abolished when the dorsal roots of the spinal nerves are severed.

Signals from the working muscles enter the spinal cord via the dorsal root; in addition, some of the fibres in the anterior root are sensory and can transmit a pressor response from the hind limb. Sensory nerve transmission in the dorsal horn may involve somatostatin and substance P [neurokinin (NK)-1] but not NK-A or NK-B [5]. It is modulated by vasopressin, and morphine attenuates this reflex, possibly by suppressing the release of substance P. The neurons are influenced by α_2-adrenergic attenuation from supraspinal centres, which could make central command signals operate on a spinal level. The exercise pressor reflex is pre-

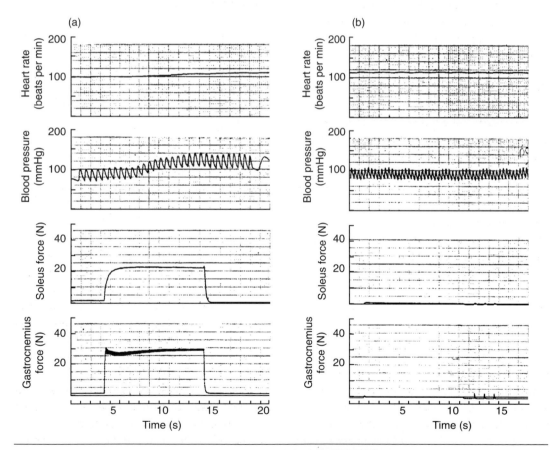

Fig. 1. Cardiovascular response to stimulation of the hind limb of a cat

Heart rate, blood pressure and force developed in the soleus and gastrocnemius muscles followed during a contraction developed after stimulation of the distal ends of the cut ventral roots (L7–S1). Control contraction (a) and contraction after neuromuscular blockade (b). Reproduced with permission from [4]. ©1987, The Physiological Society.

sent in decerebrate animals and, traditionally, afferent input is thought to reach brain stem structures, whereas spinal reflexes are operative only after long-term spinal injury. Of special interest is the finding that the baroreceptor neurons in the nucleus tractus solitarius are inhibited by peritoneal nerve stimulation [by way of γ-aminobutyric acid (GABA)] [6].

During investigation of the neural hypothesis for cardiovascular regulation in human exercise, it has become clear that influences operate very differently during the two main modalities of physical exercise, i.e. isometric (static) exercise and so-called concentric (dynamic) exercise. Furthermore, cardiovascular regulation is different during the onset of exercise and during continued exertion. In real life, the two forms of exercise are often combined, such as during weight-lifting and rowing; consequently, cardiovascular regulation becomes more complex.

Onset of static exercise

Fig. 1 shows experimental data from animal experiments in which muscle contractions were induced by stimulation of the distal ends of the cut ventral roots. The experiments demonstrated that the heart rate and blood pressure responses to a muscle contraction are delayed by approximately 1–2 s in this preparation. This delay may indicate that some signal has to be developed in the contracting muscle before the cardiovascular manifestation. Yet, when an effort is made to preserve vagal tone in the heart, the cardiovascular responses are seen earlier, arguing in favour of a significant role of mechanoreceptors.

The idea that the initial increase in heart rate is generated by central command is supported by the fact that, in humans, it is not affected by regional anaesthesia of the exercising limb(s) and is delayed by one heart beat when a muscle contraction is evoked by electrical stimulation of the muscle. Conversely, an immediate increase in heart rate is developed during evoked contraction when nerve (rather than muscle) is stimulated, which probably reflects direct activation of afferent fibres that are not easily reached when stimulation is

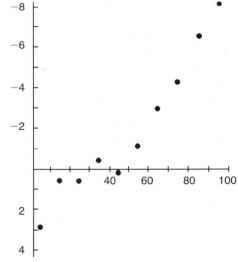

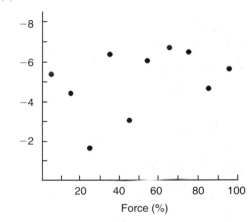

Fig. 2. Heart rate at the onset of exercise
Percentage change in electrocardiogram R–R interval at the onset of submaximal and maximal finger and hand muscle contractions (a) and maximal contractions weakened by ischaemic fatigue, tubocurarine and decamethonium (b) related to the force developed. Reproduced with permission from [7]. ©1985, The Physiological Society.

applied at a distance from the fibres. Indeed, in support of the attribution it has been observed that in humans the increase in heart rate in the first heart beat after the onset of exercise is coupled to the effort made by the subject, and not to the force generated by the muscle (Fig. 2). In fact, the onset of very-low-effort static exercise may even be associated with a decrease in heart rate. The increase in heart rate at the onset of exercise has been shown to be due to vagal withdrawal since it is blocked by atropine and left unaffected by propranolol.

Especially short-lasting static contractions are performed with a concomitant Valsalva-like manoeuvre [8], a forceful exhalation which increases blood pressure markedly but does not affect heart rate [9]. Even when every effort is made *not* to make a Valsalva-like manoeuvre — by maintaining normal breathing — blood pressure increases by approx. 10 mmHg at the onset of such exercise. It can also be argued that the part of the increase in blood pressure that is not caused by a Valsalva-like manoeuvre reflects a manifestation of vagal withdrawal. Blood pressure increases by approx. 10 mmHg after the administration of atropine. Thus the increases in both heart rate and blood pressure might be attributed to an effect of central command.

The theory that the immediate increase in blood pressure at the onset of exercise is an effect of the increase in heart rate is supported by the finding that the increase in muscle sympathetic nerve activity is delayed. However, sympathetic activity to the skin [10], and to the kidneys of animals, increases immediately at the onset of exercise, or in its anticipation. Furthermore, even muscle sympathetic nerve activity can be conditioned to 'the will' to perform intense intermittent isometric exercise, in that it is synchronized to the small force generated after partial neuromuscular blockade or regional anaesthesia [11].

Onset of dynamic exercise

Heart rate also increases immediately at the onset of dynamic exercise and, as with the onset of static exercise, the increase is often delayed by one heart beat [2], although this is not always reported. Thus the latent period for the central command influence on heart rate is calculated as being <300 ms, whereas the influence from the leg muscles takes approximately 600 ms to take effect. As at the onset of static exercise, a Valsalva manoeuvre does not affect the immediate heart rate response, which is caused by vagal withdrawal, as during static exercise.

In contrast with static exercise, the mean blood pressure response to dynamic exercise may be delayed for as much as half a minute [12]. The delay in response time is dependent on work rate; as work rate is increased, the blood pressure and heart rate responses become larger and the rise in blood pressure is manifested within a few seconds. Thus, during dynamic exercise, mean blood pressure is finely balanced between the increase in cardiac output and the almost equal decrease in total peripheral resistance. Since the decrease in total peripheral resistance follows a fixed rate related to refilling of the blood vessels in the muscles after each contraction, the initial fluctuations in blood pressure depend on how fast the increase in cardiac output becomes manifest.

Dynamic exercise with partial neuromuscular blockade lowers blood pressure for several seconds, which may be seen as evidence of central command having an effect on total peripheral vascular resistance by way of muscle blood flow. In fact, muscle blood flow increases in anticipation of exercise, and especially so with training. One explanation may be that during the initial stages of exercise, muscle blood flow is controlled under the influence of the nerves, while local metabolic factors take over with some delay. If so, adrenaline release from the adrenal glands and storage in the sympathetic nerve endings may be involved and, at least in the rat, β-receptors contribute to the initial muscle hyperaemia. Whether or not cholinergic vasodilatation exists in humans is controversial, but the idea has been raised that vasodilatation may develop in response to 'overflow' of acetylcholine from the neuromuscular transmission.

Static exercise

Early studies in humans have indicated that the cardiovascular response to static exercise depends on the relative intensity of the exercise (Fig. 2) rather than on the muscle mass involved [8]. However, from the results of more-detailed studies it can be seen that the muscle mass does have an independent influence. Sympathetic nerve activity increases not only with contraction intensity but also with muscle mass, although the increase in cardiovascular variables and sympathetic nerve activity is not simply additive of the responses manifest during contractions of single muscle groups. Furthermore, the effect of muscle mass on heart rate is manifest within the first heart beat after the onset of exercise [7] and is, therefore, likely to be due to a central as well as a peripheral mechanism. Only when contractions are carried out to exhaustion does the effect of muscle mass on cardiovascular variables disappear, thereby supporting a dominant role of central command.

Effects on the circulation ascribed to central command are easily demonstrated during human static exercise where muscles have been pharmacologically weakened by partial neuromuscular blockade. Most often the same *absolute* contraction intensity is maintained with weakened muscles; to achieve this, the effort is enhanced but the contraction maintains the same change in the interior milieu of the muscle. Conversely, by maintaining the same *relative* work load, the effort to maintain the contraction is close to constant; however, owing to the reduced contraction intensity in the weakened muscle, the signal generated by the muscle is considered to be lower.

At the same absolute work load, contractions with weakened muscles cause heart rate and blood pressure to increase more than during control contractions, and the increases are greatest when partial neuromuscular blockade is so severe that it is impossible for the subject to maintain a given contraction intensity (Fig. 3). Furthermore, during complete neuromuscular blockade, even attempted contractions are associated with a cardiovascular response. Conversely, at the same relative work load, exercise with partial neuromuscular blockade elicits a normal cardiovascular response.

The central command effect on blood pressure is due primarily to an elevated cardiac output rather than to an effect on total peripheral resistance. In the study by Gandevia and Hobbs [14] using regional anaesthesia, central command had a more obvious effect on heart rate than on blood pressure. Thus muscle sympathetic nerve activity is dominated by signals

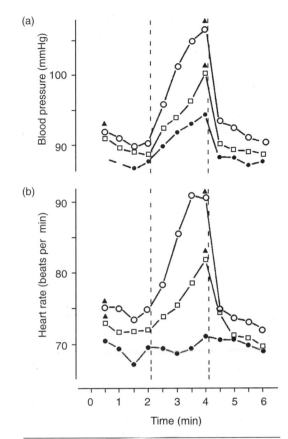

Fig. 3. Mean arterial blood pressure (a) and heart rate (b) followed at rest, during a 2 min sustained handgrip contraction corresponding to 15% of the initial maximal contraction strength and for the 2 min following recovery

● , *Control contraction;* □, *contraction with tubocurarine which the subjects were able to maintain;* ○ , *contraction with tubocurarine which the subjects were unable to maintain throughout the 2 min contraction;* ▲ , *resting values and exercise responses obtained with tubocurarine are different. Dashed lines demarcate the 2 min contraction period. Reproduced with permission from [13].* ©1989, The Physiological Society.

generated in working muscle, and central command can elevate it only marginally. In contrast, skin sympathetic nerve activity increases even in anticipation of exercise [10] and a similar central command signal appears to be responsible for constriction of the relatively large radialis and dorsalis pedis arteries.

The enhanced cardiovascular response to static exercise with weakened muscle may depend on spinal integration or, more probably, it may be the result of subjects having trained and having a current 'memory' of the required output. This is exemplified in paraplegics, whose attempts to contract their paralysed legs do not affect heart rate or blood pressure, whereas the normal responses are elicited during attempted contraction of the arm blocked by regional anaesthesia.

Thus the two neural control mechanisms may not be independent. During (dynamic) exercise with partial neuromuscular blockade, plasma adrenocorticotropin (ACTH) becomes elevated, whereas the plasma level of β-endorphin is unaffected [15]. Yet the increase in plasma concentration of both hormones is abolished when leg exercise is performed with epidural anaesthesia. This means that the apparent central command effect on corticotropin release depends on neural feedback from the working muscles.

The importance of the exercise pressor reflex during static exercise is demonstrated by normal cardiovascular (and ventilatory) responses to electrically evoked contractions. Furthermore, a normal blood pressure response to voluntary exercise at a given absolute force, and reduced responses to exercise with epidural anaesthesia at a given relative force, may be taken to indicate the importance of the exercise pressor reflex.

The blood pressure response to static exercise is not easily blocked [13]. Atropine increases resting heart rate to 100–110 beats per min, but it does not block the blood pressure response to exercise. Equally, propranolol decreases resting heart rate, and α-adrenergic blockade lowers the resting blood pressure but not the exercise response. Only with combined α- and β-adrenergic blockade is the exercise response abolished. The increase in blood pressure is primarily caused by cardiac output, although with β-adrenergic blockade it is caused by an increase in total peripheral resistance. This indicates that, although elevated, blood pressure is the primary regulated variable.

Dynamic exercise

It has been demonstrated repeatedly that electrically evoked dynamic exercise causes the same circulatory and ventilatory responses as voluntary exercise. A demonstration in humans of the muscle pressor reflex for cardiovascular regulation during voluntary dynamic exercise is shown by the increase in heart rate and blood pressure during cycling with occluded circulation to the legs. More recently, it was found that, during dynamic exercise, partial neuromuscular blockade causes heart rate and blood pressure to decrease in proportion to the reduction in work capacity (Fig. 4), i.e. in contrast to static exercise.

Essential to the hypothesis of reflex activation of the circulation is the finding that light epidural anaesthesia diminishes the blood pressure response to voluntary exercise. The abolition of the blood pressure response to evoked contractions during paralysing epidural anaesthesia [17] has confirmed this finding, despite reports to the contrary in some subjects of a pilot study. In the study by Freund et al. [18], voluntary exercise was performed with very weak muscles after epidural anaesthesia. In that situation, exercise simulates intermittent isometric contractions on the pedals of the cycle ergometer and during static exercise, epidural anaesthesia causes a normal or elevated pressure response.

During dynamic exercise with partial neuromuscular blockade, plasma catecholamines are elevated with no circulatory consequence [16], and when the blood pressure response to exercise is reduced or eliminated with epidural anaesthesia, this takes place with normal catecholamine levels [17]. Plasma catecholamines rise dramatically during maximal exercise [19], but seem to be of little circulatory consequence. During exercise, plasma catecholamines may be regarded as overspill from activation of the sympathetic nervous system. In addition, their role in substrate mobilization during exercise is unclear, but they may be important for clearance of plasma potassium.

In contrast with blood pressure, epidural anaesthesia has no effect on heart rate and ventilation during dynamic exercise [17]. For heart

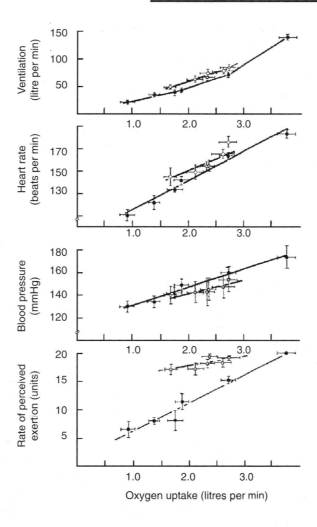

Fig. 4. Ventilation, heart rate, arterial blood pressure and rate of perceived exertion related to oxygen uptake

○, Maximal exercise during waning neuromuscular blockade with tubocurarine; ●, submaximal and maximal exercise without neuromuscular blockade. Values are given as means ± SEM for ten subjects. Reproduced with permission from [16]. ©1987, The Physiological Society.

rate, this may be due to an effect of blood temperature on the heart and activation of cardiopulmonary receptors as the central blood volume becomes enlarged, but experimental data are lacking.

The balance between the importance of central command and the muscle pressor reflex appears to be different during dynamic exercise with a small muscle group. In the studies by Davies and Sargeant [20] and Klausen et al. [21], one-leg training (of both legs) caused the expected decrease in heart rate at a given work rate (Fig. 5). However, when the subjects performed normal cycling with both legs, there appeared to be no training effect on heart rate, despite the fact that both legs were trained (independently). Innes et al. [22] also studied one-leg exercise and demonstrated a larger increase in cardiovascular variables when a weak leg was exercising than during control exercise with the healthy contralateral leg. An effect of central command on heart rate during

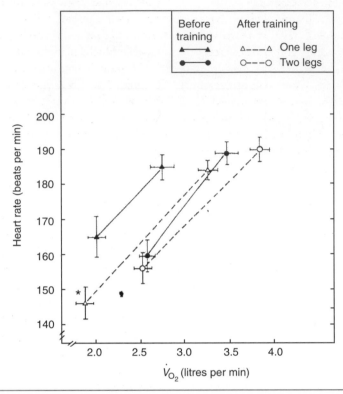

Fig. 5. Heart rate during sub-maximal and maximal one- or two-leg exercise before and after one-leg training

$\dot{V}_{O_2}$, oxygen uptake. Values are means $\pm$ SEM for six subjects. *, Post-training value different from pre-training value. Reproduced with permission from [21]. ©1982, The American Physiological Society.

dynamic exercise is also demonstrated during hypnosis, by suggesting that the workload was either light or heavy, and by learning followed by on-line display of the heart rate to the subjects to secure feedback.

Arterial occlusion

A further indication that the exercise pressor reflex is more significant during dynamic than during static exercise has been given by the finding that, after dynamic exercise, blood pressure is maintained if a cuff placed proximal to the muscle is inflated to a suprasystolic pressure before the end of exercise. In contrast, cuff inflation cannot maintain the exercise-induced elevation of blood pressure as effectively after static exercise. After both static and dynamic exercise, an arterial cuff seldom maintains an elevated heart rate.

The maintained elevation in blood pressure after dynamic exercise resulting from inflation of an arterial cuff depends on afferent nerves from the working muscle, and is reduced or eliminated by regional anaesthesia.

The argument derived from application of an arterial cuff *after* exercise for control of the circulation *during* exercise is weakened by the fact that the elevated blood pressure during exercise is caused by the increase in cardiac output, whereas it is maintained by an arterial cuff because of an increase in total peripheral resistance. It should also be noted that wearing an arterial cuff after exercise can cause pain; in that situation, the ability of the muscle afferents to affect myocardial contractility becomes manifest and blood pressure is elevated by way of cardiac output.

Activation of autonomic activity

It is a matter of speculation as to which events in the exercising muscle cause the increase in sympathetic nerve activity and cardiovascular variables. Mechanoreceptors originating in muscle spindles and the Golgi tendon organs are not likely to be involved in cardiovascular control. They transmit their signals in group-I and -II afferents that are known not to have circulatory importance. Furthermore, muscle vibration does not affect cardiovascular variables in humans. In contrast, activation of the group-III and -IV afferents are associated with significant cardiovascular responses. The group-III fibres may be more associated with connective tissue structures, whereas the group-IV fibres represent free nerve endings in the muscle [22a]. This means that the group-III afferents are more-closely associated with movement, whereas the group-IV fibres may be sensors of chemical events in the muscle.

Activation of muscle afferent nerves involves integration of many different signals. Yet the circulatory consequences depend more on pH changes in the working muscle than on a hypoxic stimulus, and the effect of potassium on muscle afferent fibres is only transient. McArdle patients (myophosphorylase deficiency), for example, show no increase in sympathetic nerve activity during exercise, since muscle [H$^+$] remains stable because no lactate is produced; in contrast, the response to other stimuli that increase sympathetic nerve traffic (e.g. a Valsalva manoeuvre and a cold pressor test) is normal [23]. Also, these studies demonstrated no association between an increase in sympathetic nerve activity and vasoconstriction, and the changes in the intracellular concentrations of phosphocreatine, inorganic phosphate or ADP — as detected by nuclear magnetic resonance spectroscopy, but diprotonated phosphate may play a role. Compared with a resting muscle pH of 7.1, it has been argued that the muscle chemoreflex has a threshold around an intracellular pH of 6.9.

Formation of lactate and, in turn, an increase in muscle [H$^+$] may be related to the involvement of fast-twitch muscle fibres. The importance of muscle fibre involvement for cardiovascular regulation during exercise was originally suggested by Petrofsky and Lind [24] from experiments involving selective neuromuscular blockade in the cat.

In animals, cardiovascular responses may be coupled to liberation of prostaglandins; in humans, indomethacin does not modify the response. Furthermore, 5-hydroxytryptamine (5-HT; serotonin) may be involved in activation of unmyelinated nerve fibres, but a 5-HT antagonist does not significantly influence cardiovascular control during dynamic exercise in humans. The increased muscle pressure during muscle contractions has an independent, blood-pressure-raising effect. Evidence is also accumulating that the interstitial volume of the muscles has an influence on the blood pressure response to exercise. Furthermore, the rise in muscle temperature during exercise increases the activity of muscle afferent nerves and the elevated blood temperature supports an increase in heart rate.

Baroreceptors

It has been considered that the cardiovascular adaptation to exercise is caused by the arterial baroreceptors. This consideration can only apply to the onset of light dynamic exercise with a large muscle mass, because blood pressure increases immediately at the onset of static exercise with or without the development of a Valsalva-like manoeuvre, and almost immediately at the onset of intense dynamic exercise. Yet arterial baroreceptors do have a modifying influence on cardiovascular control during exercise in response to elevated mean arterial and pulse pressure, and they are essential for the redistribution of cardiac output. Further activation of the arterial baroreceptors by application of pulsative negative pressure to a neck collar reduces the blood pressure and heart rate responses and also leg vascular conductance during exercise. Conversely, if the increase in blood pressure is prevented by infusion of nitroprusside, the increases in heart rate and sympathetic activity are augmented. Considered in this way, arterial baroreceptors are 'reset' to a higher pressure, but operate normally during exercise. Thus signals from both the central nervous system and the working muscles operate to control blood pressure via the influence on neural integration of the signals from the baroreceptors.

Cardiopulmonary afferent fibres are of two types related to the degree of myelinization. Myelinated cardiopulmonary afferents have effects similar to those of the arterial baroreceptors; unmyelinated fibres are associated with the 'Bezold–Jarish reflex', increasing vagal activity and decreasing sympathetic outflow. It can be speculated that activation of myelinated cardiopulmonary afferents becomes manifest during exercise due to the increased central blood volume. Thus upright exercise is associated with a decrease in muscle sympathetic nerve activity, presumably because the central venous pressure and blood volume become elevated.

Unmyelinated cardiac afferents are activated by mechanoreceptors in the left ventricle. During central volume depletion, reduction in left ventricular volume by approximately 10–20% appears to be sufficient to activate the receptors, resulting in relative bradycardia and hypotension [24a]. With the much larger change in left ventricular volumes during exercise, such receptors should be activated intensively. This is indicated by vagal activity during exercise, as reflected in the plasma concentration of the hormone pancreatic polypeptide and the heart rate variability, and it may be 'unmasked' after acclimatization to extreme altitude [25]. Evidence for the role of heart receptors for cardiovascular control during exercise has also come from the study of heart-transplant recipients. In these patients, forearm vascular resistance increased during static exercise, but decreased in control subjects. In contrast, lower body negative pressure does not affect the sympathetic nerve response to isometric contractions in normal humans.

Exercise is characterized by neither bradycardia nor hypotension. One explanation for this could be that strong sympathetic activation from the exercising muscle and also from the central nervous system over-rides the Bezold–Jarish reflex [24a]. However, the situation may not always be so clear. By intense acti-

vation of central receptors, as seen during exercise in patients with aortic stenosis, relative hypotension and bradycardia may be elicited. The importance of the central and arterial baroreceptors during exercise is underlined by the finding that, when both types of receptor are lost in animals, blood pressure decreases on transition to dynamic exercise, and recovers only gradually with time depending on the work rate.

Integration

Many of the experiments performed in order to understand the neural control of the circulation during exercise have attempted to demonstrate the importance of *either* central command *or* the exercise pressor reflex. However, the concept that the two mechanisms are mutually redundant is suggested by, for example, the fact that electrically stimulated contractions elicit the same response as does voluntary exercise. However, many experimental situations lead to results that are markedly different from the normal response. It appears that the exercise pressor reflex is able to generate the full cardiovascular response to both static and dynamic exercise, and is mainly responsible for the normal blood pressure response to dynamic exercise. Conversely, central command signals, by way of elevated heart rate and cardiac output, are able to induce more than the normal responses to static exercise and dynamic exercise with a small muscle mass. The immediate cardiovascular response to the onset of both static and dynamic exercise is likely to be generated by central command signals.

In considering cardiovascular regulation during exercise, it should be remembered that many forms of dynamic exercise are performed more or less automatically due to learning. This is the case for walking and running and, in many countries, also for cycling. Thus one significant effect of training is to reduce one's awareness of the activity and, thereby, maybe also to reduce the central command influence not only on innervation of skeletal muscle but on cardiovascular regulation as well. In support of this idea is the fact that during handgrip the increase in regional cerebral blood flow in the motor cortex is smaller with the use of the dominant as compared with the non-dominant hand. Conversely, static exercise and other unusual activities, such as one-leg exercise, represent a type of movement with which the subject is unfamiliar. Therefore, not only is the muscle contraction significantly influenced by central command signal, but so too is the cardiovascular system. It is only after repetitive practice, or training, that the control of circulation is matched to the demand of the muscle during sub-maximal dynamic exercise. During intense dynamic and static exercise this goal is never achieved, due to the increase in muscle pressure during exercise with, as a consequence, insufficient blood flow to the working muscle.

Redistribution of cardiac output during exercise

The increase in cardiac output during exercise serves the working muscles with little change in the oxygen uptake to non-exercising tissue and corresponds to approximately 5 litres of blood per min per litre of oxygen uptake. At high work rates, the increase in cardiac output tends to be somewhat smaller. Cardiac output increases at low work rates due to an increase in both heart rate and stroke volume for upright exercise. At higher work rates, stroke volume reaches a maximal value and a further increase in cardiac output becomes dependent on (a near linear) increase in heart rate.

The physiological mechanisms that increase cardiac output during exercise are not fully understood. Increased output may be coupled to some metabolic event in the exercising muscle, since it is markedly elevated during dynamic exercise in patients with, for example, McArdle's disease, demonstrating dissociation between signals controlling sympathetic nerve activity and blood flow.

During complete (surgical) epidural anaesthesia in which exercise is induced by electrical stimulation of the muscles, cardiac output is normal or only slightly reduced [17]. It could be speculated that the normal increase in cardiac output during dynamic exercise is modu-

lated around the mechanical events consequent upon activation of the venous pump. Thus the venous pump will increase preload to the heart and cardiac output will be supported by the effect of circulating catecholamines on the heart. Therefore, even cardiac transplant patients are able to increase cardiac output by way of a Starling mechanism; in these patients, heart rate also increases. Conversely, known humoral factors with vasodilatory effects can account for no more than 20% of the increase in flow that is manifest during exercise.

The classical concept is that exercise causes sympatholysis in the active musculature. However, several lines of evidence point to the fact that the sympathetic nerves do play a role in regulating blood flow to the exercising muscle. Strandell and Shephard demonstrated [26] that forearm exercise during lower body negative pressure is associated with reduced regional flow during mild work rates. During more intense exercise, the effect of lower body negative pressure became less marked.

Perhaps of more relevance to exercise is the situation when several muscle groups are working at the same time. The maximal blood flow has been reported to be as different as 40–60 ml per 100 ml of muscle per min to 200–500 ml per 100 ml of tissue per min. If the upper range is taken into consideration, it would exceed the known range of cardiac output if all muscles in the body were active and a mechanism must exist which reduces muscle dilatation to maintain blood pressure.

A reduced arm blood flow during combined arm and leg exercise was demonstrated by Secher et al. [27] as a decrease in arm venous oxygen saturation and, subsequently, a reduction in peak forearm vascular conductance has been confirmed by direct measurement of arm flow [28]. In both studies, the effect of leg exercise on arm blood flow was greatest when the load on the legs was high.

Conversely, when arm exercise is carried out with a high work rate, this may reduce leg blood flow (Fig. 6). Intense forearm exercise also reduces calf blood flow when the foot is exercising at a moderate intensity, and leg blood flow is often lower during two-leg exer-

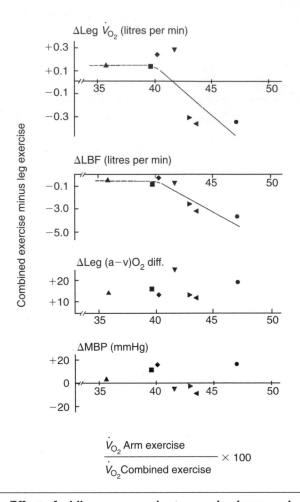

Fig. 6. Effect of adding arm exercise to ongoing leg exercise

Individual changes in leg oxygen uptake (leg $\dot{V}_{O_2}$), leg blood flow (LBF), regional arterio-venous oxygen differences [(a−v)O_2 diff.] and mean arterial blood pressure (MBP), indicate differences calculated by subtracting values obtained during separate leg exercise from values obtained during combined exercise. The abscissa is $\dot{V}_{O_2}$ for the arm load performed separately expressed as a percentage of $\dot{V}_{O_2}$ during combined exercise. Different symbols represent different subjects. Reproduced with permission from [27]. ©1977, Acta Physiologica Scandinavica.

cise as compared with one-leg. Furthermore, when cardiac output is reduced during exercise by selective β-adrenergic blockade, half of the reduction in cardiac output is manifest over the exercising legs. Indirect evidence indicates that the other half of the reduction in cardiac output, during exercise with β-blockade, affects the splanchnic blood flow. Further support for the importance of a sympathetic mechanism comes from the study of Pawelczyk et al. [29], who calculated reduced leg conductance and

increased noradrenaline overspill from the exercising legs during intense exercise.

Changes in distribution of blood volume have been evaluated by the use of technetium-labelled erythrocytes [30]. Exercise causes an increase in the thoracic blood volume which reaches 38% during maximal exercise — representing a 24% increase in heart volume and a 50% increase in lung blood volume, which is more prevailing in the upper half of the lung. Blood is mobilized from the spleen (by 46%),

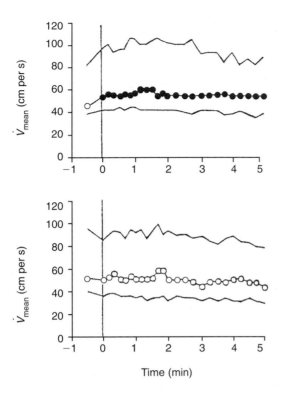

Fig. 7. Contralateral and ipsilateral middle cerebral artery mean velocity ($\dot{V}_{mean}$) during hand contractions

*Values are medians with range (upper and lower thin lines) before (-1 to 0 min) and during (0 to 5 min) exercise. Closed symbols, values significantly different from rest [31b]. Reproduced with permission from Jørgenson, G.L., et al. (1993) Am. J. Physiol. **264**, H 553–H559. ©The Physiological Society.*

kidney (by 24%) and liver (by 18%), giving an average reduction in abdominal blood volume of 19%. The reduction in the spleen blood volume is too small to account even for the relatively small increase in haematocrit observed during exercise in humans (3–4%), in comparison with some animals. In addition, the volume of blood in the legs decreases during exercise (by 23%) as a consequence of the action of the muscle venous pump.

Blood flow to non-exercising tissue

In isometric handgrip experiments, and in contrast with what is expected, blood flow has been reported to increase in the non-exercising arm due to a cholinergic or β-adrenergic mechanism. However, when muscle activity of the 'resting' arm is controlled by a display of the electromyographic activity and no activity is allowed, vasoconstriction prevails [31]. Thus the reported vasodilatation associated with unintended muscle activity may represent part of the general stress response as it is also reported during mental stress. If so, it is more prevalent for the forearm than for the calf muscles.

Skin blood flow is affected differently at the onset of exercise and when there is a need for thermoregulatory vasodilatation. The initial vasoconstriction takes place only during dynamic exercise and when a significant muscle mass is involved. This indicates that vasoconstriction is initiated by the activity of the working muscle and not by the perceived effort. This is in contrast to what is derived from the recording of skin sympathetic nerve activity [10], which then probably represents the innervation of sweat glands. Thermoregulation requires vasodilatation, but this is less during exercise at a given body temperature than at rest. It is interesting to note that tooth blood flow increases during exercise following cholinergic vasodilatation.

The splanchnic circulation is the only major vascular bed that can be manipulated to direct blood to the working muscle. By measurement of the arterio-hepatic venous oxygen difference at rest and during dynamic exercise, Clausen [31a] calculated a reduction in the splanchnic circulation in relation to heart rate. These findings demonstrated that there is a linear relationship from rest to maximal exercise corresponding with a 70% reduction at a heart rate of 200 beats per min. Accordingly, gastric mucosal pH decreases from 7.25 to 6.79.

Cerebral blood flow

It is controversial whether, in general, cerebral blood flow and oxygen uptake increase during exercise. Yet, blood flow corresponding to the large middle cerebral artery territory increases during exercise in reflection of the increase in regional cerebral blood flow to the contralateral sensorimotor area [31b].

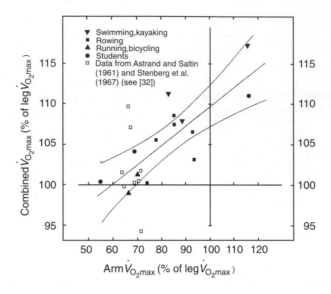

Legend (in plot):
- ▼ Swimming, kayaking
- ■ Rowing
- ▲ Running, bicycling
- ● Students
- □ Data from Astrand and Saltin (1961) and Stenberg et al. (1967) (see [32])

Y-axis: Combined $\dot{V}_{O_2max}$ (% of leg $\dot{V}_{O_2max}$)
X-axis: Arm $\dot{V}_{O_2max}$ (% of leg $\dot{V}_{O_2max}$)

Fig. 8. Maximal oxygen uptake ($\dot{V}_{O_2max}$) during combined arm plus leg exercise related to $\dot{V}_{O_2max}$ during arm cranking

Both values expressed as percentage of $\dot{V}_{O_2max}$ during leg exercise. Horizontal and vertical lines indicate situations where either combined or arm $\dot{V}_{O_2max}$ are identical to leg $\dot{V}_{O_2max}$. The 95% confidence limits for the regression line are included. Reproduced with permission from [32]. ©1974, The American Physiological Society. For references to Åstrand and Saltin, and Stenberg et al., see [32].

During rhythmic hand contractions, a small increase in regional cerebral blood flow is also manifest bilaterally deep in the brain, corresponding to the supplementary area that is presumed to represent planning of movement. It is interesting to note that the large contralateral increase in cerebral blood flow does not take place during static exercise. Also of interest is the finding that both the large contralateral and the small bilateral increase in cerebral blood flow depend on afferent input from the exercising limb, since both are eliminated by regional anaesthesia. Equally, the increase in middle cerebral artery mean-flow velocity with exercise is eliminated by regional anaesthesia of the working limb (Fig. 7).

Of the two increases in regional cerebral blood flow, the small bilateral increase is eliminated by even light regional anaesthesia of the working arm. This suggests that it reflects afferent input from unmyelinated or lightly myelinated nerve fibres associated with detection of metabolic changes in the working muscle. On the other hand, the large contralateral increase in regional cerebral blood flow, which depends on movement, is eliminated only during paralysing regional anaesthesia of the working arm. These findings suggest that this increase in regional cerebral blood flow reflects the activation of mechanoreceptors in the working limb.

The changes in regional cerebral blood flow during exercise represent integration of the signals received from the working limbs, although they do not represent 'central command'. Perhaps only a few neurons are responsible for the command signals; alternatively, these 'command neurons', which are responsible for the response to initiation and maintenance of exercise, are spread over a large part of the brain. Preliminary results on regional cerebral blood flow indicate that the insula may represent the area responsible for simultaneous increases in heart rate and blood pressure during exercise.

▶ Evidence points to blood pressure as being the primary regulated cardiovascular variable during exercise, i.e. during exercise, the increase in blood pressure may be seen as a safety measure (or overshoot) to prevent blood pressure from falling in response to the often large decrease in total peripheral resistance provided by flow to existing muscle.

▶ Flow to other regions (notably to the splanchic region) decreases, with the consequent 'mobilization' of blood from such organs.

The cardiovascular system as a limitation to maximal exercise

The classic model for exercise metabolism was formulated by Hill in 1925, who assumed that metabolism can be described as a sum of a capacitance and a rate multiplied by the duration of exercise. On the basis of these assumptions, an accurate description has been derived for both running [33] and rowing [34]. In physiological terms, the capacitance may be considered to represent the stores of oxygen in blood and muscle, high-energy phosphate available and liberation of energy through the formation of lactate. The rate of energy turnover in sustained exercise will be reflected by the rate of oxygen uptake, and Hill and Lupton [35] were the first to determine that this rate can reach a maximal level, termed maximal oxygen uptake. Only recently has an attempt been made to define the maximal value for the capacitance, termed 'the maximal oxygen deficit'.

Hill and Lupton [35] considered $\dot{V}_{O_2max}$ to represent an estimate of cardiac output, and it has been widely accepted that, in whole-body exercise involving a large muscle mass, $\dot{V}_{O_2max}$ is limited by the circulatory system, more specifically by the heart. There is little, if any, direct evidence for this notion, and other indirect evidence seems to contradict the idea that the heart limits $\dot{V}_{O_2max}$.

Blood pressure may be taken as a balance between the increase in cardiac output and the decrease in total peripheral vascular resistance.

With an elevated mean arterial pressure during exercise, it follows that cardiac output always increases more than peripheral resistance decreases. Thus $\dot{V}_{O_2max}$ is closely related to vascular conductance, and an increase in $\dot{V}_{O_2max}$ is associated with a decrease in peripheral resistance [31a]. As pointed out by Clausen, patients with arteriosclerotic heart disease can increase their work capacity as a result of endurance training. The training effect is due to reduced afterload, and exercise is stopped at the same load on the heart as that expressed by the product of heart rate and systolic blood pressure.

If it is considered that the heart is limiting $\dot{V}_{O_2max}$, it should also be considered whether it is the pumping capacity (the capacity of the heart as a muscle), or the diastolic function of the heart (the capacity for a large stroke volume), that is important. Like skeletal muscles, the heart adapts to training with the development of a 'sport heart' [36] and this is not simply due to an effect of the exercise associated with elevation of plasma adrenaline.

In weight-lifters especially, the thickness of the wall of the heart increases, whereas in runners the enlargement of the heart is due to an increase in its internal diameters. Runners have a larger $\dot{V}_{O_2max}$ than weight-lifters, and even rowers who combine large internal diameters of the heart with a large wall thickness do not possess especially large $\dot{V}_{O_2max}$ in relation to their body size [19]. On this evidence it could be argued that the pumping capacity of the heart is unlikely to limit $\dot{V}_{O_2max}$. In support of this argument it might be pointed out that healthy subjects never complain of pain in the chest during maximal exercise, whereas they may develop severe pain in their muscles.

It is evident that subjects with a large $\dot{V}_{O_2max}$ are characterized by large stroke volumes, relating the function of the heart to its diastolic function. In fact, pericardiectomy increases $\dot{V}_{O_2max}$ in the dog and in the pig. However, the heart cannot deliver more blood than it receives, arguing for the role of blood volume in $\dot{V}_{O_2max}$. It is known that blood volume increases with endurance training and also that $\dot{V}_{O_2max}$ (and work capacity) increases acutely by administration of blood (blood dop-

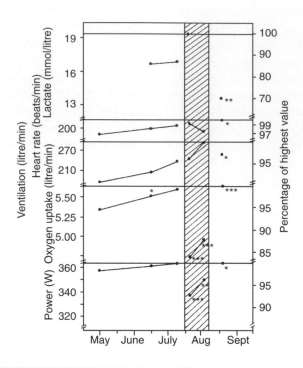

Fig. 9. Blood lactate, heart rate, pulmonary ventilation, $\dot{V}_{O_2max}$ and power on a rowing ergometer before and after training at altitude

Values are averages of nine oarsmen. In response to altitude training, the only change was a reduction of blood lactate after maximal exercise at sea level. Reproduced with permission from [19].

ing). Thus the increase in blood volume may be considered to be a relatively fast adaptation to training, whereas the changes in heart dimension take longer.

Central to the discussion of limiting factors for $\dot{V}_{O_2max}$ is a comparison of the values obtained during different types of exercise. $\dot{V}_{O_2max}$ is larger when a large muscle mass is involved in exercise. Accordingly, maximal values for oxygen uptake obtained by small muscle groups are sometimes referred to as 'peak', as opposed to values obtained during running or cycling. However, when arm exercise is added to leg exercise even higher values are measured than in the so-called *maximal* oxygen uptake measured in cycling or running.

The peak oxygen uptake achieved in exercise with a small muscle mass depends on specific training. In untrained subjects, upper body (arm) exercise elicits a $\dot{V}_{O_2max}$ which is approxi-

mately 70% of that generated during leg exercise (Fig. 8). However, in rowers the value is approximately 85%, and in swimmers a higher $\dot{V}_{O_2max}$ may be obtained during exercise with the arms than during leg exercise [32]. Theses observations indicate that in exercise involving a small muscle mass, $\dot{V}_{O_2max}$ depends more on local adaptation in the working muscle than on the capacity of the heart [20]. Both capillary density and enzymes are of importance for oxidative metabolism only in the trained muscles [36a]. Of these two factors, capillary density is the variable more closely related to changes in $\dot{V}_{O_2max}$.

Of interest, analyses of oxygen transport versus peripheral diffusion for limiting $\dot{V}_{O_2max}$ suggest that, in untrained subjects, $\dot{V}_{O_2max}$ may not be limited by oxygen delivery to the exercising muscle. After training, oxygen delivery (leg blood flow) increases. However, the increase in muscle diffusion capacity for oxygen increases even more, resulting in enlarged oxygen extraction. Also, glycogen utilization and lactate production are reduced as a consequence of training.

There is also evidence that ventilation may reach a level during exercise with large muscle groups where it becomes a problem. This is illustrated during rowing, where the respiratory muscles also have to assist the propulsive force generation, and tidal volume becomes limited as the respiratory frequency is entrained to the stroke frequency [37]. Traditionally, ventilation is not considered to be a limiting factor for the $\dot{V}_{O_2max}$, because it increases out of proportion to oxygen uptake during severe exercise. However, with training, ventilation during maximal exercise increases markedly and may be in demand to saturate arterial blood. Haemoglobin saturation is made difficult because of the decrease in pH, which may reach 6.8 with the involvement of a large muscle mass [38]. For that reason, haemoglobin saturation may be as low as 90% and further arterial desaturation (to 82%) indicates a pulmonary diffusion limitation to oxygen transport and/or that the ventilation–perfusion distribution becomes more widely dispersed at high work rates. Yet, it needs to be established at what level of venous hypoxaemia it becomes a problem to saturate the arterial blood.

▶ Even in healthy humans, intense exercise is associated with muscular pain, whereas only patients with ischaemic heart disease develop chest pain. It is therefore unlikely that the pumping capacity of the heart is normally a limitation to exercise and cardiac output increases in response to acute plasma expansion.
▶ Work capacity is increased as a consequence of an increase in the volume of red blood cells, supporting further the fact that, during exercise, cardiac output is limited by preload to the heart.

▶ Endurance training increases the internal-diameter of the heart, and the maximal stroke volume is thereby increaseed. Whether this increased stroke volume is representative of an enlarged heart, or reflects passive expansion in response to the training-induced elevation of blood volume, needs to be established but, in animals, pericardiectomy has been shown to increase cardiac output during exercise.

Altitude

At altitude, the reduced exercise capacity challenges many of the concepts derived from studies carried out at sea level. The inspired oxygen tension decreases more than that corresponding to the decrease in ambient oxygen tension due to the saturation of alveolar air with water (47 mmHg). Furthermore, the reduction in $\dot{V}_{O_2max}$ is most severe in those subjects with the lowest pulmonary diffusion capacity. Thus reduction in $\dot{V}_{O_2max}$ is directly related to the reduction in arterial saturation with oxygen.

Exercise capacity may also be limited by the reduced ability to form lactate after acclimatization to hypoxia. The reduction in the ability to form lactate is developed over a period of approximately 3 weeks. It may be related to an adrenergic mechanism, since lactate formation is abolished at altitude after administration of propranolol. In addition, the ability to form lactate after acclimatization declines in parallel with the reduction in plasma adrenaline during exercise but, in contrast to what would be expected, the reduction in blood lactate is manifest when exercise involves large rather than small muscle groups [25].

The reduced ability to form lactate remains for some time after return to sea level (Fig. 9) [19]. Yet, with unchanged $\dot{V}_{O_2max}$ after altitude training, the reduction in lactate formation does not appear to reduce work capacity over a period of 6 min.

The reduction in the ability to form lactate may also explain why maximal heart rate tends to decrease during prolonged periods at altitude. It has been demonstrated that activation of the sympathetic nervous system during exercise is closely associated with the decrease in muscle pH. Such findings make it difficult to determine whether the reduction in lactate formation triggers the sympathetic nerve activity at altitude, or vice versa.

Exercise at altitude also causes other problems. Prolonged exposure to altitude may change the dynamics of the cerebral circulation of interest with respect to the development of 'mountain sickness' associated with cerebral oedema. At sea level and during acute exposure to a simulated altitude of 4300 m, mean velocity in the internal carotid artery increases by 15–33% during exercise [39]. However, after 18 days at altitude, that there is no increase in internal artery velocity during exercise maybe reflecting a decrease in arterial carbon dioxide tension with increasing hyperventilation.

While a normal blood pressure response is maintained during exercise with acute exposure to hypoxia, acclimatization may change the response to hypotension during exercise with a large muscle mass [25]. On the other hand, when exercise is performed with only one leg, the normal pressor response is maintained. It could be speculated that exercise with a large muscle mass activates central baroreceptors after acclimatization. Alternatively, the stimulation of the sympathetic nervous system may be too weak to account for sufficient control of the circulation. The two explanations are not mutually exclusive.

▶ Altitude training remains popular among many elite athletes and 'nitrogen chambers' have been constructed such that athletes are able to sleep under conditions which simulate high altitude, even during their normal training.
▶ Although it is evident that an increased haemoglobin volume is associated with an equally increased $\dot{V}_{O_2max}$ and work capacity, there is no evidence to support the hypothesis that altitude training enhances $\dot{V}_{O_2max}$ or work capacity.

▶ On the contrary, all evidence points to altitude training as offering no advantage for sea-level performance compared with similar sea-level training.
▶ Part of the problem with altitude training is linked to reduced work capacity for endurance events, and it is suggested that 'sleeping high, training low' can combine the potential stimulatory effect due to hypoxia on erythropoiesis with the capacity to maintain the normal training haemoglobin volume; however, an effect of altitude training (or a 'pressure chamber' simulation) on the volume of red blood cells needs to be established.

Training

For the planning of training it is of utmost importance to understand what limits work capacity. If one common factor were limiting, training for most types of exercise requiring a large aerobic metabolism would not need to be 'performance task-specific'. In contrast, if local factors in the active muscles were to play a significant role, this would indicate that training ought to be based on exercising the muscle groups specific to the performance required.

It was demonstrated by Clausen [31a] that leg training decreases heart rate almost equally during arm and leg exercise. However, arm training had little effect on heart rate during leg exercise. Exercise with the trained arms or legs increased oxygen extraction over the working limbs, while oxygen extraction over non-exercising tissue and hepatic clearance became less pronounced. Leg training was found to increase cardiac output and blood pressure during arm exercise, i.e. similar to the heart-rate finding.

These results are similar to the observation previously mentioned, i.e. that one-leg training does not affect the heart rate response to two-leg exercise [20,21]. Of practical importance is the observation by Lewis et al. [40] that, although arm training increases both arm and leg performance, the increase in arm performance was greatest. Similarly, when the subjects performed leg training, the increase in work capacity with the leg was greater than with the arms. Thus training induces marked local adaptations of relevance not only to the metabolic response but also of importance for cardiovascular function. This is reflected in the increased regional vascular conductance of trained humans.

The specificity of training for $\dot{V}_{O_2max}$ was demonstrated elegantly by Holmér [41]. He followed the $\dot{V}_{O_2max}$ of a swimmer during swimming and running over 6 years. The $\dot{V}_{O_2max}$ remained unchanged during running. However, during free-style swimming and 'arm exercise', $\dot{V}_{O_2max}$ changed considerably from season to season. This means that, especially for athletes, $\dot{V}_{O_2max}$ should be measured in an ergometer designed to simulate the specific type of exercise or, even better, be evaluated during the actual task itself. This latter approach is to be recommended wherever practicable, not least because of criticism that the ergometer simulation may not be comparable with the actual performance. Even in tasks that superficially appear remarkably similar, such as cycle ergometer exercise and cycling, there may be important, and for some applications critical, differences.

▶ Many aspects of endurance training are common to different activities, such as running or cycling.

▶ Notably, most human activities that involve leg exercise and training for one event will have significant implications for other, and often very similar, events; however, when other muscle groups (notably the arms) play a significant role in exercise, the specificity of the training response becomes evident.

▶ Endurance training has not only general systemic effects, but also gives rise to significant local adaptations in the working muscle, including an increase in the number of capillaries and adaptation of the mitochondrial enzymes. Accordingly, any particular task needs a significant amount of 'specific' training.

Conclusions

▶ For endurance events, the overwhelming evidence is that performance is limited by the circulatory system, and substantial efforts have been made to determine how such limitation arises.

▶ During exercise, the circulatory system is challenged not only in its capacity to deliver blood — and thereby oxygen — to the exercising muscles, but it also needs to maintain its own integrity in order to perfuse vital organs such as the brain.

▶ Numerous neural control mechanisms have been described in order to explain how blood pressure is maintained and, most often, also elevated during exercise, despite the marked decrease in total peripheral resistance as a consequence of muscle vasodilatation.

▶ It appears that the priority of the cardiovascular system is to maintain blood pressure, even at the expense of an increased flow to the active muscle and the brain, in order that each organ receives no more blood than that which can be 'allowed' to provide for the blood pressure that the arterial baroreceptors are registered to detect.

▶ In untrained humans, a circulatory restraint on muscle flow may be irrelevant, since oxygen transport is also limited by the diffusion capacity of the muscle; however, owing to an increased number of capillaries in the muscles due to training, oxygen transport to the muscles may become the dominating factor in determining the ability of tissue to take up oxygen.

▶ For the athlete, an acute increase in blood volume allows for an increased preload to the heart and, thereby, also an increase in both cardiac output and oxygen uptake during exercise.

▶ The limiting factor for exercise is localized if the demand for an increase in cardiac output is low, but with the active engagement of a large muscle mass (and especially so in the athlete), a centralized limitation to flow has to conform to the necessity of

> maintaining blood pressure.
>
> ▶ With the involvement of many muscle groups, it becomes increasingly difficult to fully oxygenate the arterial blood.
>
> ▶ Desaturation of arterial blood of oxygen to less than 90% has been reported during running in athletes, and this is quite common in rowing, since the arterial pH is lowered and binding of oxygen to haemoglobin is consequently more difficult. In other words, the circulatory system is no longer the only transport system that limits oxygen delivery to the muscles.

Further reading

Åstrand, P.-O. and Rodahl, K. (1984) *Textbook of Work Physiology*, McGraw-Hill, New York

Rowell, L.B. (1992) *Human Cardiovascular Control*, Oxford University Press, New York

Spyer, K.M. (1994) Central nervous mechanisms contributing to cardiovascular control. *J. Physiol.* **474**, 1–19

References

1. Eldrige, F.L., Milhorn, D.E., Kiley, J.P. and Waldrop, T.G. (1985) Stimulation by central command of locomotion, respiration and circulation during exercise. *Respir. Physiol.* **59**, 313–337

2. Krogh, A. and Lindhard, J. (1913) The regulation of respiration and circulation during the initial stages of muscular work. *J. Physiol.* **47**, 112–136

3. Coote, J.H., Hilton, S.M. and Perez-Gonzalea, J.F. (1971) The reflex nature of the pressor response to muscular exercise. *J. Physiol.* **215**, 789–804

4. Iwamoto, G.A., Mitchell, J.H., Mizuno, M. and Secher, N.H. (1987) Cardiovascular responses at the onset of exercise with partial neuromuscular blockade in cat and man. *J. Physiol.* **384**, 39–47

5. Hill, J.M., Pickar, J.G. and Kaufman, M.P. (1992) Attenuation of reflex pressor and ventilatory responses to static contraction by an NK-1 receptor antagonist. *J. Appl. Physiol.* **73**, 1389–1395

6. McMahon, S.E., McWilliam, P.N., Robertson, J. and Kaye, J.C. (1992) Inhibition of carotid sinus baroreceptor neurons in the nucleus tractus solitarius of the anaesthetized cat by electrical stimulation of hindlimb afferent fibres. *J. Physiol.* **452**, 224P

7. Secher, N.H. (1985) Heart rate at the onset of static exercise in man with partial neuromuscular blockade. *J. Physiol.* **368**, 481–490

8. Lind, A.R. (1983) Cardiovascular adjustments to isometric contractions: static effort. In *Handbook of Physiology: The Cardiovascular System*, section 2, vol. III (Shephard, J.T., Abboud, F.M. and Geiger, S.R., eds.), pp. 947–966, American Physiological Society, Bethesda

9. Lassen, A., Mitchell, J.H., Reeves, Jr, D.R., Rogers, H.B. and Secher, N.H. (1989) Cardiovascular responses to static exercise in man with topical nervous blockade. *J. Physiol.* **409**, 333–341

10. Vissing, S.F., Scherrer, U. and Victor, R.G. (1991) Stimulation of skin sympathetic nerve discharge by central command. *Circ. Res.* **69**, 228–238

11. Victor, R.G., Secher, N.H., Lyon, T. and Mitchell, J.H. (1995) Central command increases muscle sympathetic nerve activity during intense intermittent isometric exercise in humans. *Circ. Res.* **76**, 127–131

12. Holmgren, A. (1956) Circulatory changes during muscular work in man: with special reference to arterial and central venous pressures in the systemic circulation. *Scand. J. Clin. Lab. Invest.* **8**, suppl 24

13. Mitchell, J.H., Reeves, Jr, D.R., Rogers, H.B., Secher, N.H. and Victor, R.G. (1989) Autonomic blockade and cardiovascular responses to static exercise in partially curarized man. *J. Physiol.* **413**, 433–445

14. Gandevia, S.C. and Hobbs, S.F. (1990) Cardiovascular responses to static exercise in man: central and reflex contributions. *J. Physiol.* **430**, 105–117

15. Kjær, M., Secher, N.H., Bach, F.W. and Galbo, H. (1987) Role of motor center activity for hormonal changes and substrate mobilization in humans. *Am. J. Physiol.* **253**, R687–R695

16. Galbo, H., Kjær, M. and Secher, N.H. (1987) Cardiovascular, ventilatory and catecholamine responses to maximal exercise in partially curarized man. *J. Physiol.* **389**, 557–568

17. Strange, S., Secher, N.H., Pawelczyk, J.A., Karpaka, J., Christensen, N.J., Mitchell, J.H. and Saltin, B. (1993) Cardiovascular responses to electrically induced dynamic exercise during epidural anaesthesia. *J. Physiol.* **470**, 693–704

18. Freund, P.R., Rowell, L.B., Murphy, T.M., Hobbs, S.F. and Butler, S.H. (1979) Blockade of the pressor response to muscle ischemia by sensory nerve block in man. *Am. J. Physiol.* **237**, H433–H439

19. Jensen, K., Nielsen, T., Fiskestrand, Å., Lund, J.O., Christensen, N.J. and Secher, N.H. (1993) Altitude training does not increase maximal oxygen uptake or work capacity. *Scand. J. Med. Sci. Sports* **3**, 256–262

20. Davies, C.M.T. and Sargeant, A.J. (1975) Effects of training on the physiological response to one- and two-leg work. *J. Appl. Physiol.* **38**, 377–381

21. Klausen, K., Secher, N.H., Clausen, J.P., Hartling, O. and Trap-Jensen, J. (1982) Central and regional circulatory adaptations to one leg training. *J. Appl. Physiol.* **52**, 976–983

22. Innes, J.A., De Cort, S.C., Evans, P.J. and Guz, A. (1992) Central command influences cardiorespiratory response to dynamic exercise in humans with unilateral weakness. *J. Physiol.* **448**, 551–563

22a. Mitchell, J.H. and Schmidt, R.F. (1983) Cardiovascular reflex control by afferent fibers from skeletal muscle receptors. In *Handbook of Physiology: The Cardiovascular System*, section 2, vol. III, (Shephard, J.T., Abboud, F.M. and Geiger, S.R., eds.), pp 623–658, American Physiological Society, Bethesda

23. Pryor, S.L., Lewis, S.F., Haller, R.G., Bertocci, L.A. and Victor, R.G. (1990) Impairment of sympathetic activation during static exercise in patients with muscle phosphorylase deficiency (McArdle's Disease) *J. Clin. Invest.* **85**, 1444–1449

24. Petrofsky, J.S. and Lind, A.R. (1980) The blood pressure response during isometric exercise in fast and slow twitch skeletal muscle in the cat. *Eur. J. Appl. Physiol.* **44**, 223–230

24a. Secher, N.H., Jacobsen, J., Friedman, D.B. and Matzen, S. (1992) Bradycardia during reversible hypovolaemic shock: associated neural reflex mechanisms and clinical implications. *Clin. Exp. Pharmacol. Physiol.* **19**, 733–743

25. Savard, G.K., Areskog, N.-H. and Saltin, B. (1995) Cardiovascular response to exercise in humans following acclimatization to extreme altitude. *Acta Physiol. Scand.* **154**, 499–509

26. Strandell, T. and Shephard, J.T. (1967) The effect in humans on increased sympathetic activity on the blood flow to active muscles. *Acta Med. Scand.* **472**, 146–167

27. Secher, N.H., Clausen, J.P., Klausen, K., Noer, I. and Trap–Jensen, J. (1977) Central and regional circulatory effects of adding arm exercise to leg exercise. *Acta Physiol. Scand.* **100**, 288–297

28. Sinoway, L. and Prophet, S. (1990) Skeletal muscle metaboreceptor stimulation opposes peak metabolic vasodilatation in humans. *Circ. Res.* **66**, 1576–1584

29. Pawelczyk, J.A., Hanel, B., Pawelczyk, R.A., Warberg, J. and Secher, N.H. (1992) Leg vasoconstriction during dynamic exercise with reduced cardiac output. *J. Appl. Physiol.* **73**, 1838–1846

30. Flamm, S.D., Taki, J., Moore, R., Lewis, S.F., Keech, F., Maltais, F., Ahmad, M., Callahan, R., Dragotakes, S., Alpert, N. and Strauss, H.W. (1990) Redistribution of regional and organ blood volume and effect on cardiac function in relation to upright exercise intensity in healthy human subjects. *Circulation* **81**, 1550–1559

31. Cotzias, C. and Marshall, J.M. (1993) Vascular and electromyographic responses evoked in forearm muscle by isometric contraction of the contralateral forearm. *Clin. Autonom. Res.* **3**, 21–30

31a. Clausen, J.P. (1976) Circulatory adjustments to dynamic exercise and effect of physical training in normal subjects and in patients with coronary artery disease. *Progr. Cardiovasc. Disease* **28**, 459–495

31b. Jørgensen, G.L. (1995) Transcranial doppler ultrasound for cerebral perfusion. *Acta Physiol Scand* **154**, Suppl. 625, pp. 1–43

32. Secher, N.H., Ruberg-Larsen, N., Brinkhorst, R.A. and Bonde-Petersen, F. (1974) Maximal oxygen uptake during arm cranking and combined arm plus leg exercise. *J. Appl. Physiol.* **36**, 515–518

33. Lloyd, B.B. (1966) The energetics of running: an analysis of World records. *Adv. Sci.* **22**, 515–530

34. Secher, N.H. (1983) The physiology of rowing. *J. Sports Sci.* **1**, 23–53

35. Hill, A.V. and Lupton, H. (1923) Muscular exercise, lactate acid, and the supply and utilization of oxygen. *Q. J. Med.* **16**, 135–171

36. Secher, K. (1923) Experimentelle Untersuchungen über die Grösse des Herzens nach einem Aufhören des Trainerens. *Zeitschrift für die gesamte experimentelle Medizin* **22**, 290–295

36a. Saltin, B. and Gollnick, P.D. (1983) Skeletal muscle adaptability, significance for metabolism and performance. In *Handbook of Physiology: Skeletal Muscle*, section 10 (Peachey, L.D., Adrian, R.H. and Geiger, S.R., eds.), pp. 555–631, American Physiological Society, Bethesda

37. Steinacker, J.M., Both, M. and Whipp, B.J. (1993) Pulmonary mechanics and entertainment of respiration and stroke rate during rowing. *Int. J. Sports Med.* **14(1)**, S15–S19

38. Hanel, B., Clifford, P.S. and Secher, N.H. (1994) Restricted postexercise pulmonary diffusion capacity does not impair maximal transport for O_2. *J. Appl. Physiol.* **77**, 2408–2412

39. Huang, S.Y., Tawney, K.W., Bender, P.R., Grover, B.M., McCullough, R.E., McCullough, R.G., Micco, A.J., Manco-Johnson, M., Cymerman, A., Green, E.R. and Reeves, J.T. (1991) Internal carotid flow velocity with exercise before and after acclimatization to 4,300 m. *J. Appl. Physiol.* **71**, 1469–1476

40. Lewis, S., Thompson, P., Areskog, N.-H., Vodak, P., Marconyak, M., DeBusk, R., Mellan, S. and Haskell, W. (1980) Transfer effects of endurance training to exercise with untrained limbs. *Eur. J. Appl. Physiol.* **44**, 25–34

41. Holmér, I. (1974) Physiology of Swimming Man. *Acta Physiol. Scand.* **407**, 1–55

8

Determinants and limitations of pulmonary gas exchange during exercise

Susan A. Ward

Centre for Exercise Science and Medicine, University of Glasgow, Glasgow G12 8QQ, U.K.

Introduction

The demands of increased muscle metabolism during exercise require rapid cardiovascular and ventilatory system responses to minimize disturbances of blood and tissue acid–base homeostasis. Consequently, limitations of these system responses — with respect not only to their magnitude but also to their dynamic features — will in turn establish the operating range for the transport and exchange of oxygen and CO_2.

The preceding chapters have addressed the control of intramuscular energetics during exercise; these have set the stage for a systematic analysis of the factors that limit the rate of aerobic metabolism and thus impact on physical performance. The contributions of cardio-circulatory mechanisms and of intramuscular tissue transport are addressed in detail in Chapter 7 and Chapter 3, respectively, whereas pulmonary limitations — both mechanical and vascular — are considered in Chapter 9. This chapter aims to provide a frame of reference for consideration of muscular and cardio-respiratory system limitations to pulmonary gas exchange, exercise tolerance and physical performance.

It is important to recognize that both the demands for pulmonary gas exchange and the actual response profiles that ensue are characteristic of the intensity of the task: i.e. although the maximum O_2 uptake ($\mu\dot{V}_{O_2}$) is an important parameter for determining the tolerable range of work rates, this range can also be usefully partitioned by the lactate threshold (θ_L) [1] into two different intensity domains, 'moderate' and 'high'.

θ_L is usually defined as the threshold work rate (or, more appropriately, O_2 uptake) at which a sustained metabolic (predominantly lactic) acidaemia results [1]. Therefore, the range of work rates within which there is no sustained metabolic acidaemia may be considered to be of moderate intensity. This intensity domain is characterized by the attainment of steady-states of pulmonary gas exchange ($\dot{V}_{O_2}$, $\dot{V}_{CO_2}$), cardiac output ($\dot{Q}_T$) and ventilation ($\dot{V}_E$); these work rates can, therefore, be sustained comfortably for prolonged periods. Work rates in excess of θ_L can be regarded as being of high intensity, since they are characterized by metabolic acidaemia and a more-rapid onset of fatigue. The acquisition of a steady state, when it can be attained, is substantially delayed.

Oxygen exchange

The maximum rate of O_2 utilization at the cytochrome oxidase terminus of the mitochondrial electron-transport chain sets the limit for muscle O_2 consumption ($\dot{Q}_{MO_2}$) and, therefore, pulmonary O_2 uptake ($\dot{V}_{O_2}$) (for example, see [2]). It is, therefore, not surprising that $\mu\dot{V}_{O_2}$ is dependent upon the muscle mass involved ([3–5]; see Chapter 7 in this volume). For example, $\mu\dot{V}_{O_2}$ for leg exercise is appreciably higher than for arm exercise. The highest recorded values for $\mu\dot{V}_{O_2}$ (in excess of 80 ml per kg of body weight per min) are typically ascribed to elite cross-country skiers, who utilize both upper- and lower-body muscle groups. However, in the face of progressive muscle recruitment beyond ~50% of the total muscle mass, $\mu\dot{V}_{O_2}$ eventually reaches a maximum for the body as a whole ($\mu\dot{V}_{T_{O_2}}$). Further recruitment of muscle groups is considered not to result in any further increase in $\mu\dot{V}_{O_2}$ (see [4] for discussion).

For the majority of athletic tasks, however, it appears that $\mu\dot{V}_{O_2}$ does not approach $\mu\dot{V}_{T_{O_2}}$. For these situations, $\mu\dot{V}_{O_2}$ measured for a particular exercise mode is traditionally regarded as the physiological limit for prolonged performance at that exercise mode. It is thus important to distinguish $\mu\dot{V}_{O_2}$ from a symptom-limited or peak $\dot{V}_{O_2}$ which, although representing a maximum for a particular task, cannot be taken as unequivocal evidence of a system limitation to tissue, and therefore pulmonary-gas exchange (see [4–6] for discussion).

The limitations to $\mu\dot{V}_{O_2}$ have come to be addressed in terms of a cascade of mechanisms that extend from the mouth to the mitochondria in the working muscles [4–8]. This involves transfer of O_2 both by convective flow (in the airways and the blood) and by diffusion (across the alveolar–capillary membrane in the lung,

and between the capillary bed and the mitochondrial matrix in skeletal muscle) (Fig. 1). This approach typically considers the exchange process within the constraint of the 'steady-state'. However, steady-state conditions are not attained instantaneously, even for moderate work rates; and for high work rates, steady-states rarely prevail. Considerations of the limitations to whole-body O_2 (and CO_2) transfer should, therefore, incorporate these dynamic response features, and their control.

Moderate-intensity exercise

The ability to transport and exchange O_2 is not compromised at moderate-intensity work rates. Rather, limitations to athletic performance in this intensity domain (i.e. in endurance events such as the marathon, ultramarathon and full-course triathlon) involve factors such as substrate availability, thermoregulation and hydration status [3].

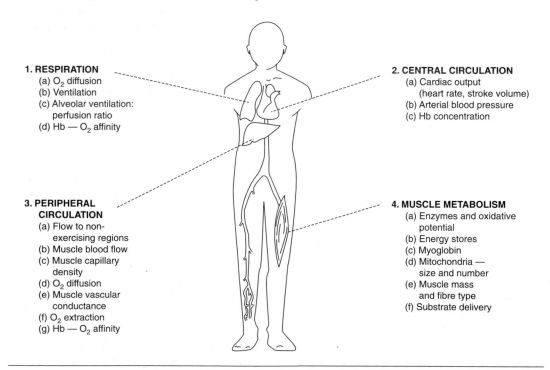

Fig. 1. Potential sites of limitation for O_2 exchange and utilization
Abbreviation used: Hb, haemoglobin. From Human Circulation: Regulation During Physical Stress by Loring, B. Rowell. Copyright ©1986 by Oxford University Press, Inc. Used by permission of Oxford University Press, Inc.

Muscle O_2 consumption

The increase in muscle O_2 utilization during exercise can usefully be considered in terms of the Fick principle; i.e.

$$\dot{Q}_{MO_2} = \dot{Q}_M \cdot (C_{aO_2} - C_{MvO_2}) \qquad (1)$$

where C_{aO_2} and C_{MvO_2} are the O_2 contents of arterial and muscle venous blood, respectively, and $\dot{Q}_M$ is muscle blood flow.

Increases in $\dot{Q}_{MO_2}$ are accomplished by: (i) an increased rate of vascular O_2 delivery, attributable almost entirely to $\dot{Q}_M$ [i.e. since arterial blood is almost completely saturated with O_2 at rest (97–98%), it cannot, therefore, undergo any appreciable further increase during exercise]; and (ii) increased rates of O_2 extraction, reflected in a declining C_{MvO_2}.

$\dot{Q}_M$ increases in proportion to $\dot{Q}_{MO_2}$ during progressive exercise [4,9,10]; values of 20 litres per min or so may be attained by healthy active subjects, and 40 litres per min, or more, in elite endurance athletes [4,10]. Both $\dot{Q}_T$ and the degree to which it is preferentially diverted to the exercising muscles contribute to the $\dot{Q}_M$ response [4,9,10].

In humans, $\dot{Q}_{MO_2}$ during moderate exercise is thought to increase towards a new steady-state in an essentially mono-exponential fashion following the onset of moderate, constant-load exercise:

$$\Delta\dot{Q}_{MO_2}(t) = \Delta\dot{Q}_{MO_2}(ss) \cdot (1 - e^{-t/\tau}) \qquad (2)$$

where $\Delta\dot{Q}_{MO_2}(ss)$ is the steady-state $\dot{Q}_{MO_2}$ increment above baseline, $\Delta\dot{Q}_{MO_2}(t)$ is the $\dot{Q}_{MO_2}$ increment at time t, and τ is the time constant of the $\dot{Q}_{MO_2}$ response. Direct corroboration of this assertion is not straightforward in humans, however. Although $\dot{Q}_{O_2}$ can be measured across an active limb (by the Fick principle; see, for example, Chapter 3 in this volume), this approach is invasive, does not necessarily 'capture' all the involved musculature and is limited by its ability to discriminate rapid-response dynamics rigorously (i.e. with time constants of the order of 0.5 to 1.0 min). However, the temporal pattern of the phosphocreatine (PCr) response in human muscle (estimated by ^{31}P-NMRS [11–14]) and its coupled exponentiality with $\dot{Q}_{MO_2}$ in other species (for example, see [15]) supports the assertion that the $\dot{Q}_{MO_2}$ response in humans is mono-exponential during moderate exercise.

$\dot{Q}_{MO_2}$ is widely regarded to be triggered by increased rates of free-energy utilization, resulting from faster rates of high-energy phosphate-bond hydrolysis — with muscle perfusion being adequate or even excessive for the current demands. Although the precise mechanisms subserving muscle respiratory-control kinetics remain conjectural, various control models have been proposed. These draw on work in experimental animals and, more recently, in humans (see earlier) and include: (i) creatine-kinase-catalysed PCr kinetics, induced by changes in creatine (Cr) levels; (ii) ADP levels; and (iii) the 'phosphorylation potential', defined as $\ln\{[ATP]/[ADP]\cdot[P_i]\}$ [15–18].

In contrast, an alternative view argues for a vascular limitation of $\dot{Q}_{MO_2}$ and $\dot{V}_{O_2}$ kinetics during moderate exercise. According to the proponents of this control scheme, it is the dynamics of the arterial O_2 delivery to the muscle that govern $\dot{Q}_{MO_2}$ [19]. The evidence that has been cited in support of this contention derives from:

- the reported slowing of $\dot{V}_{O_2}$ kinetics against a background of β-adrenergic blockade and during supine rather than upright exercise — conditions considered to also slow $\dot{Q}_M$ kinetics; and
- the restoration to upright levels of the slowed $\dot{V}_{O_2}$ kinetics for supine exercise by concurrent application of lower-body negative pressure, consistent with a 'normalization' of the $\dot{Q}_M$ response.

However, there is also evidence suggesting that $\dot{Q}_{MO_2}$ and $\dot{V}_{O_2}$ kinetics are not perfusion-limited during moderate exercise. The imposition of lower-body positive pressure to levels that are known to reduce $\dot{Q}_M$ has been demonstrated not to slow $\dot{V}_{O_2}$ kinetics [20]. In addition, $\dot{V}_{O_2}$ dynamics have been shown to be unaffected by a bout of prior high-intensity exercise [21]. These authors argued that the residual effect of the metabolic acidaemia (and possibly other metabolic sequelae) would induce vasodilatation, thereby speeding the dynamics of $\dot{Q}_M$ at the onset of exercise. In addition, Doppler-based measurements of $\dot{Q}_T$ [22] and femoral artery blood flow [23] yield response dynamics that are considerably faster than for $\dot{V}_{O_2}$; this degree of dynamic discrepancy does not support O_2 delivery being a critical determinant of $\dot{V}_{O_2}$ kinetics.

Pulmonary O_2 uptake

The expression of $\dot{Q}_{MO_2}$ response profiles at the lungs is influenced by the vascular transit time from the exercising muscles (T_{tr}) and the change in the magnitude of the intervening O_2 stores (see [24,25] for further discussion). T_{tr} will clearly depend on the volume of the intervening vascular pool ($\dot{Q}_v$) and the flow through it ($\dot{Q}_v$), i.e. $T_{tr} = \dot{Q}_v/\dot{Q}_v$ or the capacitance to conductance ratio. T_{tr} has been estimated to be of the order of 15–20 s for moderate-intensity cycle ergometry. The presence of this transit delay induces a short period following exercise onset (phase 1) in which the mixed venous blood entering the pulmonary capillary circulation does not reflect the increased metabolic demands of the exercising muscles. The increase in $\dot{V}_{O_2}$ that is typically observed in the first few breaths of exercise is ascribable, therefore, to the simultaneous increase in pulmonary blood flow ($\dot{Q}_p$) that occurs at this time. That is (by the Fick principle) as:

$$\dot{V}_{O_2} = \dot{Q}_P \cdot (C_{aO_2} - C_{\bar{v}O_2}) \qquad (3)$$

where $C_{\bar{v}O_2}$ is the mixed venous O_2 content, then:

$$\dot{V}_{O_2} = \dot{Q}_P \cdot k \qquad (4)$$

where k is the constant that reflects the caloric equivalent of the $\dot{V}_{O_2}$. C_{aO_2} and $C_{\bar{v}O_2}$ are assumed not to change appreciably during phase 1. It has been demonstrated that during cycle ergometry, performed in the upright position, $C_{\bar{v}O_2}$ does not change during phase 1 if the

work-rate increment is preceded by even the lightest attainable work rate (i.e. unloaded cycling) or when the work-rate increment is imposed from rest in the supine position. However, Casaburi et al. [26] have demonstrated a rapid fall of $C_{\bar{v}O_2}$ in the first few seconds of exercise performed from rest in the upright position. This has been interpreted to reflect the admixture within the venous return of blood from a region with a high resting ratio of $\dot{Q}_{O_2}$ to perfusion. It means, therefore, that although the phase-1 $\dot{V}_{O_2}$ response may be dominated by the $\dot{Q}_p$ mechanism, it may not be exclusively so.

Following this period of 'cardiodynamic' gas exchange, $\dot{V}_{O_2}$ increases mono-exponentially towards its steady-state (phase 2) [19,24,25, 27,28] (Fig. 2 and Fig. 3):

$$\Delta \dot{V}_{O_2}(t) = \Delta \dot{V}_{O_2}(ss) \cdot (1 - e^{-(\tau - \delta/t)}) \qquad (5)$$

where $\Delta \dot{V}_{O_2}(ss)$ is the steady-state $\dot{V}_{O_2}$ increment above baseline, $\Delta \dot{V}_{O_2}(t)$ is the increment at time t, and δ is a delay term reflecting the muscle-to-lung transit time. $\tau \dot{V}_{O_2}$ is of the order of 30–40 s in healthy young individuals; as the response is exponential, this provides a steady state within about 2–3 min [i.e. in a period of four time constants, $\dot{V}_{O_2}(t)$ will have approached to within 98% of $\dot{V}_{O_2}(ss)$]. However, $\tau \dot{V}_{O_2}$ tends to be shorter in trained subjects [30], but appreciably longer in elderly, sedentary subjects [31].

These rapid $\tau \dot{V}_{O_2}$ kinetics result in a relatively small O_2 deficit (O_2Def) at these work rates (see [24,25] for further discussion). That is:

$$O_2Def = \Delta \dot{V}_{O_2}(ss) \cdot \tau' \qquad (6)$$

where τ' is the 'effective' $\dot{V}_{O_2}$ time constant (or 'mean response time'). Interestingly, $\Delta \dot{V}_{O_2}(ss)$ increases as a linear function of work rate ($\dot{W}$) in the moderate-intensity domain (Fig. 4), with a slope parameter ($\Delta \dot{V}_{O_2}(ss)/\Delta \dot{W}$) that is relatively invariant of fitness and is of the order of 10 ml per min per W for cycle ergometry (for example, see [32]).

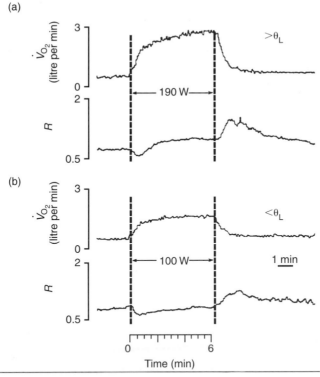

Fig 2. Breath-by-breath responses of pulmonary O_2 uptake ($\dot{V}_{O_2}$) and the respiratory gas exchange ratio (R) to a single 6 min bout of constant-load cycling

Vertical lines indicate onset and cessation of exercise. (a) Above θ_L (190 W); (b) below θ_L (100 W). Note the symmetry of the on- and off-transients of $\dot{V}_{O_2}$ below θ_L. Above θ_L, the on-transient of $\dot{V}_{O_2}$ was much slower than the off-transient. After a transient decline in R (reflecting increased rates of CO_2 stores wash-in early in the exercise), there was a subsequent increase in R beyond 1 because of subsequent stores wash-out via bicarbonate-buffering of lactic acid and respiratory compensation for the acidaemia for the supra-θ_L, but not the sub-θ_L test. Modified with permission from [29]. ©1991, The Physiological Society.

It is, therefore, possible to estimate the work efficiency (η) (see [32,33] for further discussion):

$$\eta = \frac{\Delta \dot{W}}{\Delta \dot{V}_{O_2} \cdot k} \qquad (7)$$

This, in turn, depends on the metabolic respiratory quotient ($RQ = \dot{Q}_{CO_2}/\dot{Q}_{O_2}$), which most commonly is estimated as the respiratory gas exchange ratio ($R = \dot{V}_{CO_2}/\dot{V}_{O_2}$). It is, therefore, essential that steady-state conditions prevail.

This can be problematic at high work rates because of the substantially longer time required to attain new steady states of $\dot{V}_{CO_2}$ and $\dot{V}_{O_2}$. It is important to recognize that $\Delta \dot{W}$ in this formulation represents the rate of working beyond that required simply to move the limbs: for cycling, this is the W increment *above* unloaded cycling (or 0 W), and $\Delta \dot{V}_{O_2}$ is the corresponding $\dot{V}_{O_2}$ response. As a result, in an obese individual or an athlete with highly developed leg musculature, the O_2 cost of unloaded cycling will be abnormally high (in

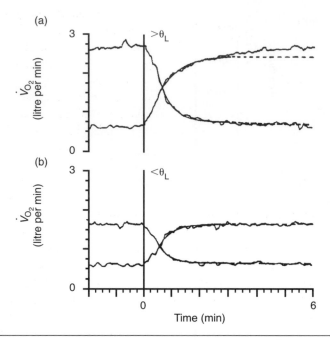

Fig. 3.   Averaged, breath-by-breath responses of $\dot{V}_{O_2}$ to a 6 min bout of constant-load cycling (4 repetitions)

For ease of comparison, the on- and off-transients are overlaid; superimposed on these is the best-fit mono-exponential over the first 3 min of exercise and recovery (solid lines), with the dashed lines indicating the extension of the mono-exponential to the 6 min point of the work bout. (a) Above θ_L (210 W); (b) below θ_L (110 W). Note that the symmetrical sub-θ_L $\dot{V}_{O_2}$ responses conform to a mono-exponential function. This was also the case for the off-transient $\dot{V}_{O_2}$ response above θ_L; however, the on-transient response showed evidence of a delayed and slowly developing 'excess' component between 3 and 6 mins. Reproduced with permission from [29]. ©1991, The Physiological Society.

proportion to the mass of the legs); however, the slope of the $\Delta\dot{V}_{O_2}$–$\Delta\dot{W}$ relationship, and therefore η, will be normal at ~30%.

While the resolution of these kinetic parameters of the $\dot{V}_{O_2}$ response may seem, at first sight, simply to be a technical exercise, they do provide insight into the fundamental determinants of exercise energetics. This is exemplified by the following expression (derived from eqns. 6 and 7) which inter-relates four of the primary parameters defining aerobic performance [34]:

$$\frac{O_2\mathrm{Def}}{\Delta\dot{W}} = \frac{\tau}{\eta} \qquad (8)$$

The extent to which the phase-2 $\dot{V}_{O_2}$ response is influenced by the intervening O_2 stores in its transformation from the $\dot{Q}_{MO_2}$ response remains uncertain (see [24,25] for further discussion). Although the O_2 storage capacity is far less than that for CO_2, there are nonetheless changes that occur in the volume of stored O_2 during exercise (see Chapter 3 in this volume). For example, there is:

• a reduction of tissue P_{O_2} in the contracting units;

• a greater extraction of O_2 from the perfusing blood, reflected in a decrease in muscle

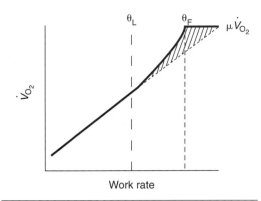

Fig. 4. Schematic representation of the relationship between $\dot{V}_{O_2}$ and work rate ($\dot{W}$) over the tolerable range of constant-load work rates, performed to the steady-state for the moderate ($<\theta_L$) and heavy (between θ_L and θ_F) ranges and to the limit of tolerance for the severe ($>\theta_F$) range

Note the relationship is linear $<\theta_L$, steepens progressively between θ_L and θ_F, attaining a maximum (i.e. $\mu\dot{V}_{O_2}$) for all higher work rates. The cross-hatched region indicates the contribution from the 'excess' component.

venous and subsequently mixed venous O_2 contents; and

• possibly a small contribution of oxymyoglobin desaturation in 'oxidative' slow-twitch fibres (however, as the P_{50}, i.e. the P_{O_2} at which 50% of the myoglobin is oxygenated, for this reaction is ~5 mmHg [6], the degree of O_2 unloading is likely to be insignificant for moderate work rates).

Cerretelli and Di Prampero [27] have argued that recruitment of the stored O_2 would necessitate $\tau\dot{V}_{O_2}$ being longer than $\tau\dot{Q}_{MO_2}$. However, based on a modelling analysis that incorporated not only considerations of the stores themselves but also the responses of $\dot{Q}_M$ and venous return, Barstow et al. [35] have proposed that, despite the utilization of stored O_2, $\tau\dot{V}_{O_2}$ is likely to reflect that of $\dot{Q}_{MO_2}$ closely (i.e. to within $\leq10\%$). This is supported by the recent results of Grassi et al. [36], utilizing direct measurement of $\dot{Q}_{MO_2}$.

A major factor impeding the resolution of these issues in humans has been the inability to establish precisely the dynamic inter-relationships between the proposed control components of the muscles' high-energy phosphate pool, muscle O_2 utilization and pulmonary O_2 uptake. This requires monitoring with a sufficient temporal density to allow unequivocal descriptions of response dynamics to be undertaken. For this reason, therefore, muscle biopsy approaches have proved of limited value. However, ^{31}P-NMRS has provided considerable insight into muscle respiratory control. Several investigators have reported that [PCr] declines in a mono-exponential fashion during constant-load exercise in humans [11–14]. Furthermore, it has been reported that $\tau\dot{V}_{O_2}$ and τ[PCr] are similar [13,37]. However, it is necessary to recognize the technical shortcomings of such studies. For example, because of: (i) different exercise modalities being used for the two profiles (and on different occasions) and (ii) the increase in muscle metabolic rate being so small (because of the relatively small muscle mass that has typically been used — e.g. forearm, calf), satisfactory estimation of kinetic parameters remains a concern. However, it is now possible to make simultaneous measurements of the dynamic high-energy-phosphate profiles and pulmonary O_2-exchange kinetics with a large-enough muscle mass for confident parameter estimation [14].

▶ For moderate-intensity exercise, increases in muscle O_2 consumption ($\dot{Q}_{MO_2}$) are accomplished by increases in muscle perfusion and O_2 extraction.

▶ $\dot{Q}_{MO_2}$ responds as a mono-exponential to moderate, constant-load exercise, and is probably controlled by factors related to high-energy phosphate turnover.

▶ The resulting pulmonary O_2 uptake ($\dot{V}_{O_2}$) response is also mono-exponential (there is also an initial, brief cardiodynamic component).

Lactate threshold

Despite there being widespread agreement for the existence of a threshold for muscle and blood [lactate] increase, the underlying mechanism(s) remain the source of considerable debate. (There are investigators who dispute this 'threshold' behaviour, however [38,39].) Three primary causes have been proposed and are considered briefly here (see [1,33,40–43] for further discussion).

Limitation of O_2 availability

This hypothesis proposes that the onset of lactic acidosis reflects a point at which the rate at which O_2 is made available to the terminal oxidant in the electron-transport chain can no longer keep pace with the requirement for ATP hydrolysis. The additional ATP requirement is, therefore, met from the anaerobic catabolism of carbohydrate.

Support for this proposal derives largely from demonstrations that procedures designed to increase O_2 delivery to the working muscle (e.g. addition of O_2 to the inspired air) increase θ_L and lead to a smaller increase in blood [lactate] at a given supra-θ_L work rate. In contrast, strategies that induce tissue hypoxia lead to a lowering of θ_L and a higher blood [lactate] at a given supra-θ_L work rate. These include: acute reduction of inspired P_{O_2} (hypoxic hypoxia); reduced perfusion of the working muscles (stagnant hypoxia); reduced levels of available

haemoglobin resulting, for example, from anaemia or low-dose inhalation of carbon monoxide (anaemic hypoxia); and mitochondrial electron-transfer abnormalities (histotoxic hypoxia). It should be noted, however, that hyperoxia may not necessarily improve O_2 delivery to the working muscles (at least to the same extent); i.e. there may well be off-setting effects of the hyperoxia on vascular perfusion (lower cardiac output, a smaller exercise-related reduction of vascular resistance in the working muscles). Likewise, the tendency for O_2 delivery to be reduced by lowering arterial P_{O_2} (P_{aO_2}) may be counteracted by a higher $\dot{Q}_M$ (because of the lower P_{O_2}).

Furthermore, lactate formation does not necessarily depend on a limitation to O_2 availability. Lactate is produced when glycolysis proceeds aerobically, in a proportion to pyruvate production that is dictated by the equilibrium constant of lactate dehydrogenase (i.e. in accordance with the law of mass action). Blood lactate levels can, therefore, increase simply in response to a stimulation of aerobic glycolysis. They will also increase if the overall rate of lactate production exceeds its rate of utilization.

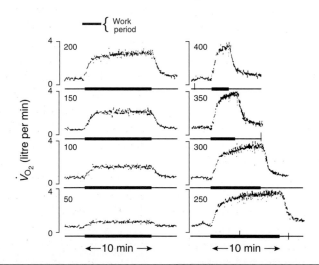

Fig. 5. **Breath-by-breath responses of $\dot{V}_{O_2}$ to single bouts of constant-load cycling performed for 10 min or to the limit of tolerance (50, 100, 150, 200, 250, 300, 350 and 400 W)**

The thick horizontal bars indicate the exercise time. Note that a steady state could no longer be reached when the work rate exceeded 150 W; rather, $\dot{V}_{O_2}$ continued to rise, reaching the same level (i.e. $\mu\dot{V}_{O_2}$) at the end of each higher work bout. Modified with permission from [45]. ©1980, Academic Press, New York.

Enzymic rate limitation

This hypothesis postulates that certain oxidative enzymes, such as succinate dehydrogenase in the Krebs cycle and cytochrome oxidase in the electron-transport chain, become rate limiting at θ_L. Evidence cited in support of this proposition is that both θ_L and muscle oxidative capacity can be increased by endurance training. However, there is as yet no evidence of mitochondrial enzymic rate limitation at θ_L. Indeed, the demonstration that θ_L can be increased by raising the inspired P_{O_2} is not consistent with limitations of oxidative enzyme activity in the presence of an adequate level of tissue oxygenation P_{O_2} inducing the increased [lactate].

Muscle fibre type

This hypothesis proposes that the fibre-type recruitment pattern mediates θ_L. It is well-known that, because of their high oxidative capacity, slow-twitch muscle fibres confer the ability to sustain work rates aerobically. However, these fibres have limited power-generating capabilities, and the generation of high-power outputs, therefore, requires the additional recruitment of fast-twitch fibres, whose low oxidative capacity and high glycolytic capacity will result in increased rates of anaerobic glycolysis and increased lactate production. This mechanism, it should be noted, can operate even if an excess of O_2 is available to a muscle cell.

In summary, although each of these proposals has its proponents and opponents, their individual contributions (if any) to the [lactate] profiles underlying θ_L are at present uncertain. This reflects the interpretational limitations imposed by the technological approaches that are currently available for use in humans. There appears to be little justification, for example, for any *a priori* assumption that profiles of P_{O_2} and [lactate] measured in muscle venous blood are representative of the muscle itself. The heterogeneities that have been demonstrated with respect to fibre type, perfusion and metabolic rate in a wide range of muscles means that some fibres may be operating at a P_{O_2} significantly less (and, therefore, producing lactate at significantly higher rates) than the mean value represented in the muscle venous effluent (see [4,42] for further discussion). An additional complication is the inability, in humans, to sample venous blood from an appropriate site. Femoral venous blood, for example, has been sampled during cycle-ergometer exercise. However, blood from this site not only includes contributions from relatively inactive muscle groups (e.g. in the feet), it also does not 'see' contributions from the most-rostral regions of the quadriceps femoris or the gluteal muscles. Despite these limitations, θ_L provides a useful parameter of interest when addressing limitations to performance in the high-intensity domain.

> ▶ The lactate threshold (θ_L) is an important demarcator of the moderate and heavy domains of exercise.
> ▶ Three factors are variously proposed to account for this threshold behaviour: limitation of O_2 availability, enzymic rate limitation and muscle fibre type recruitment.

High-intensity exercise

Athletic events for which the energy requirement cannot be met solely from aerobic means rely also on anaerobic sources of ATP production. This rapidly depletes the available glycogen stores and also leads to a sustained lactic acidosis, predisposing to fatigue. The tolerable duration of a task in this supra-θ_L domain is, therefore, relatively short.

The $\dot{V}_{O_2}$ response profile is more complex than for moderate exercise, as a result of superimposition of a slow kinetic component on to the 'basic' underlying mono-exponential process (see [24,25,44] for further discussion; Fig. 2, Fig. 3 and Fig. 5). There is, however, a range of supra-θ_L work rates in which a delayed steady-state of $\dot{V}_{O_2}$ may be reached ('heavy-intensity'). At the work rates in this intensity domain, the associated increases in arterial [lactate] and [H+] can also be stabilized. However, it is important to note that $\dot{V}_{O_2}$ is caused to increase to values greater than those predicted from purely aerobic steady-state demands; i.e.

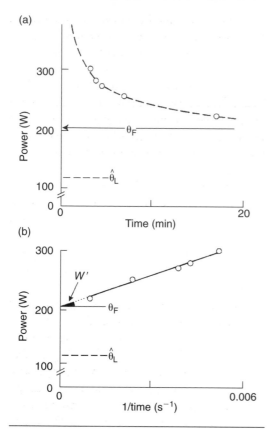

Fig. 6. Estimation of Q_F and W'

(a) The curvilinear relationship between power (or work rate) and its tolerable duration in a normal subject. (b) The linear transform of the power–duration relationship — i.e. power vs. 1/time; this relationship extrapolates to the fatigue threshold (θ_F). θ_L is the estimated lactate threshold, using standard pulmonary gas exchange criteria [33]. Modified with permission from [48]. ©1988, Taylor & Francis. http://www.tandf.co.uk/journals/erg.htm.

rather than the characteristic 10 ml per min·W for moderate exercise (see above), values of 13 ml per min·W are not uncommon during tests of 10–15 min duration (Fig. 4). This component may, therefore, be termed 'excess' O_2 uptake [$\dot{V}_{O_2}$(xs)]. As a result, the previously linear $\dot{V}_{O_2}$–$\dot{W}$ relationship becomes progressively steeper in this intensity domain, if the slow $\dot{V}_{O_2}$(xs) component is allowed to develop (Fig. 4).

Heavy exercise

The $\dot{V}_{O_2}$ kinetics in this intensity domain have been demonstrated to be asymmetrical, in con-

trast with moderate exercise (see [24,25] for further discussion; Fig. 1–Fig. 3). Whereas at least two components are required to characterize the on-transient [i.e. the phase-2 and $\dot{V}_{O_2}$(xs) components], the off-transient response often remains mono-exponential (Fig. 2 and Fig. 3). This has been argued to rule out a significant involvement of factors related to the O_2 cost of respiratory and cardiac work and to the Q_{10} effect (i.e. the temperature coefficient which quantifies the influence of a 10°C rise in temperature on a reaction or response) as mediators of $\dot{V}_{O_2}$(xs), as these would be expected to influence the off- as well as the on-transient response.

The delayed onset of the $\dot{V}_{O_2}$(xs) component seems to preclude rigorous estimation of the O_2 deficit above θ_L (eqn. 6) (see [25] for further discussion). That is, a summation is required of the O_2 deficits of both the initial phase-2 component and the delayed $\dot{V}_{O_2}$(xs) component. Although it is feasible to estimate both the effective τ and the asymptotic value of the phase-2 component (for example, see [25,44]), there are uncertainties associated with characterization of $\dot{V}_{O_2}$(xs). For example, when is $\dot{V}_{O_2}$(xs) first induced after exercise onset? And what is it that triggers $\dot{V}_{O_2}$(xs)? If the trigger is a single control mechanism with an abrupt onset, then an exponential-like response for $\dot{V}_{O_2}$(xs) would be a reasonable first assumption. In contrast, a more-complex control mechanism having several, serially recruited elements might also yield a $\dot{V}_{O_2}$(xs) response that resembles an exponential process. While the response could be adequately modelled by an exponential process [46,47], it could never accurately represent the physiological control process. Under these circumstances, there can be no meaningful physiological equivalent to the single τ parameter deriving from such an exercise. As these uncertainties cannot be resolved at present, considerations of the O_2 deficit at supra-θ_L work rates should be made with extreme caution.

Evidence currently available suggests that the highest $\dot{V}_{O_2}$ at which a steady-state can be attained coincides with the highest work rate at which arterial pH does not continue to fall during the course of the work, and arterial [lactate] does not continue to rise (i.e. the highest sustainable blood [lactate], averaging ~4–5 mM but with considerable individual variability) [48]. This $\dot{V}_{O_2}$ has also been shown to correspond to the asymptote of the power-duration curve — also termed the 'fatigue threshold' (θ_F) or 'critical power' — and typically occurs about halfway between θ_L and $\mu\dot{V}_{O_2}$ [48] (Fig. 6).

Severe exercise

At even higher work rates, steady-states are unattainable ('severe-intensity'). Rather, $\dot{V}_{O_2}$ continues to increase until $\mu\dot{V}_{O_2}$ is attained and, thereafter, fatigue rapidly ensues (see [24,25] for further discussion; Fig. 2 and Fig. 3). It appears that the higher the work rate, the more rapidly the $\dot{V}_{O_2}$(xs) component projects towards $\mu\dot{V}_{O_2}$ and the shorter the tolerable duration (t) of the task will be (Fig. 5). This feature is likely to be a significant contributor to the form of the power–duration curve for high-intensity exercise which, for cycle ergometry, has been shown to be hyperbolic [48] (Fig. 6); i.e.

$$(\dot{W} - \theta_F) \cdot t = W' \quad \text{or} \quad \dot{W} = W'/t + \theta_F \quad (9)$$

where t is the tolerable duration, and W' is the curvature constant. The parameter W' represents a constant amount of work which can be performed above θ_L, presumably supported by a particular constant pool of energy [48]. The tolerable duration (t) would, therefore, be dictated by the rate at which this pool is depleted: rapidly for high work rates; slower for lower work rates.

At these work rates, a further complexity is introduced into the $\dot{V}_{O_2} - \dot{W}$ relationship [24,25, 48] (Fig. 4): a discontinuity results at θ_F. That is, all work rates performed to the limit of tolerance above θ_F induce $\mu\dot{V}_{O_2}$; in other words, the $\dot{V}_{O_2}$ response necessarily levels off as a function of further increases in $\dot{W}$. This means that the $\dot{V}_{O_2}-\dot{W}$ relationship for a series of constant-load fatiguing work bouts does not retain the simple 'linearity' demonstrable for more moderate exercise (Fig. 4).

There are several unavoidable implications of these complex $\dot{V}_{O_2}$ kinetics above θ_L [24,25]:
(i) Steady states of $\dot{V}_{O_2}$ cannot be attained above θ_F.
(ii) V_{O_2} is no longer a linear function of $\dot{W}$. That is, the $\dot{V}_{O_2}$(xs) component (when the work bout is of sufficient duration for its expression) causes the $\dot{V}_{O_2}-\dot{W}$ relationship to steepen, and then to attain a maximum for all fatiguing work rates above θ_F.
(iii) The physiological basis of the power–duration curve having a single aerobic term (with a rapid τ of ~10–20 s) is not warranted (see [48] for further discussion).
(iv) The conventional means of computing the O_2 deficit is inappropriate in this intensity domain.
(v) The concept of the maximum accumulated O_2 deficit [49] appears questionable.

Furthermore, as there is a wide inter-subject variability in the fraction of $\mu\dot{V}_{O_2}$ that the lactate and fatigue thresholds represent, there seems to be little justification for continuing the practice of assigning exercise intensity solely in terms of either a multiple of the resting metabolic rate ('mets') or as a percentage of $\mu\dot{V}_{O_2}$ [25]. Neither do there appear to be grounds for discriminating between 'sub-maximal' and 'maximal' for high-intensity work rates that lie above θ_F; all work rates in this intensity domain become 'maximal' — i.e. they all induce $\mu\dot{V}_{O_2}$ at fatigue.

However, it is precisely because the $\dot{V}_{O_2}$(xs) component is both slow and of delayed onset that its influence is virtually undetectable during rapid-incremental tests of the ramp or small-step kind [32,33] and also for constant-load tests in which the subject reaches $\mu\dot{V}_{O_2}$ in only a few minutes [25]. When the incrementation rate is slow, however, the $\dot{V}_{O_2}$(xs) component does become evident as an upward concavity in the $\dot{V}_{O_2}-\dot{W}$ relationship (Fig. 4).

Little is known about the control of $\dot{Q}_{MO_2}$ and $\dot{V}_{O_2}$ in high-intensity exercise. This reflects technical constraints related to monitoring of muscle O_2 exchange and high-energy phosphate profile. In addition, there is a procedural limitation relating to the necessary number of multiple repeats of the exercise protocol that should be completed to secure an adequate signal-to-noise ratio for discrimination of kinetic parameters; i.e. because of the fatiguing nature of such tests, each test should be performed on a different day. Furthermore, there is the added challenge of discriminating the parameters of the excess component.

The extent to which the phase-2 component of the $\dot{V}_{O_2}$ kinetics at these work rates is controlled by muscle respiratory control mechanisms awaits appropriately designed studies that combine, for example, ^{31}P-NMRS and pulmonary gas-exchange monitoring. It has been reported that the phase-2 $\tau\dot{V}_{O_2}$ is similar to that for sub-θ_L exercise [44], suggestive of control exerted by muscle respiration. Others, however, have described $\tau\dot{V}_{O_2}$ to be slower for supra-θ_L work rates, concluding this to reflect an inadequate O_2 utilization (presumably because of a relative perfusion inadequacy to the contractile elements) that results in a metabolic acidosis to support the energy demands (see [25] for further discussion).

The few ^{31}P-NMRS investigations that have focused on the high-intensity domain are more revealing with regard to the control of the slow component of the $\dot{V}_{O_2}$ kinetics. For example, Binzone et al. [50] have described a profile of intramuscular [PCr] decline during high-intensity constant-load exercise that continued throughout the work bout, rather than stabilizing as was the case for more-moderate exercise. Using near-infrared spectroscopy, a similarly progressive O_2 desaturation in the quadriceps femoris muscles has been demonstrated during high-intensity constant-load cycling that parallels the slow excess component of the simultaneously monitored $\dot{V}_{O_2}$ response [51,52]. Also, Poole et al. [53] have concluded that $\dot{Q}_{O_2}$ measured across the lower limb manifests a progressive slow component which contributes up to 80% of that seen in $\dot{V}_{O_2}$. Interestingly, the magnitude of $\dot{V}_{O_2}$(xs) has been reported to be less prominent for subjects having a high proportion of type 1 fibres in the vastus lateralis [54]. Taken together, these diverse observations suggest that an important control element of $\dot{V}_{O_2}$ dynamics originates within the working muscles at these high work rates.

▶ At work rates between θ_L and the fatigue threshold (θ_F), i.e. heavy intensity, the V_{O_2} response is more complex, with a delayed, slowly developing 'excess' component being superimposed on the underlying mono-exponential component.

▶ Above θ_F, V_{O_2} does not attain a steady state but rises inexorably towards the maximum V_{O_2} along a trajectory set by the 'excess' component (severe intensity).

Limitations to O_2 utilization

The preceding discussion has indicated that exercise intensity and, more specifically, the lactate and fatigue threshold parameters are important in determining the kinetic behaviour of $\dot{V}_{O_2}$. Whether or not $\dot{V}_{O_2}$ attains maximal values for a particular exercise modality, such as cycle-ergometer exercise, will depend on whether the demands of the task take $\dot{V}_{O_2}$ beyond θ_F. Kinetic considerations dictate that $\dot{V}_{O_2}$ can be brought to its maximum (or peak) only above θ_F but, importantly, for the entire supra-θ_F range.

The limitations to $\dot{V}_{O_2}$ are usually addressed by consideration of the O_2 exchange along a sequence of structural components, whose operation is described through the physical laws governing convective and diffusive mechanisms of O_2 transfer (i.e. Fick's principle and Fick's law, respectively) [4–8]. The reader is also referred to Rowell's insightful analysis of this problem [4] (Fig. 1) and the contribution of Wagner [5]. The key elements are summarized below.

Pulmonary transfer

In exercising subjects with normal pulmonary function, there is no evidence extant to suggest any significant limitation to the convective transfer of O_2 from the atmosphere to the level of the alveolar ducts — where the 'front' of inspired gas is thought to become 'stationary' (see [6]). Neither is diffusion from this site across the alveolar interior to the alveolar-capillary membrane likely to be limiting during exercise; i.e. the P_{O_2} gradient and alveolar dimensions appear to allow for sufficient rates of O_2

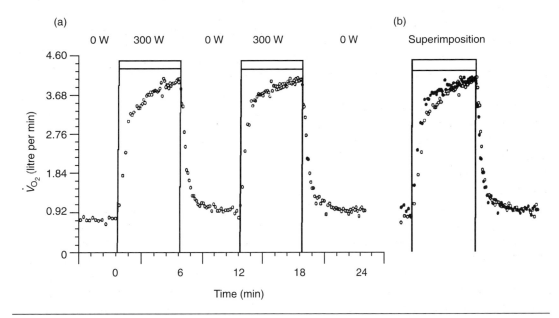

Fig. 7. Speeding of supra-θ_L $\dot{V}_{O_2}$ kinetics

(a) Response of $\dot{V}_{O_2}$ to two consecutive 6-min bouts of constant-load supra-θ_L cycling, separated by 6 min of unloaded cycling. (b) Superimposition of $\dot{V}_{O_2}$ response from first work bout ($\circ$) and that from second work bout ($\bullet$). Note that the $\dot{V}_{O_2}$ response to the second work bout develops more rapidly than for the first. Reproduced with permission from [21]. ©1996, The American Physiological Society.

transfer well within the time available for exchange (the breath duration). Consequently, alveolar P_{O_2} (P_{AO_2}) remains relatively stable at resting values for moderate work rates, and actually increases progressively above θ_L. This, as discussed below, reflects a ventilatory response out of proportion to $\dot{V}_{O_2}$ to clear additional, non-metabolic CO_2 liberated from the buffering of lactic acid and to provide compensation for the low arterial pH (see, for example [32,33]).

Traditionally, the transfer of O_2 from alveolar gas to the pulmonary circulation has not been regarded as a limiting factor to $\dot{V}_{O_2}$ during exercise. This is usually based on the stability of P_{aO_2} throughout the entire tolerable work-rate range. However, as discussed in greater detail in Chapter 9 in this volume and by Johnson et al. [6], two observations suggest that this may not be the case, certainly at high work rates:

• the widening of the alveolar-to-arterial P_{O_2} difference (i.e. while P_{aO_2} may not actually fall, it fails to 'keep pace' with the rising P_{AO_2}), and
• the presence, in some individuals, of arterial desaturation.

These effects may involve:
• a diffusion impairment across the alveolar–capillary membrane (i.e. with pulmonary end-capillary P_{O_2} failing to attain alveolar values) — this is consequent to the high cardiac outputs which reduce pulmonary capillary transit times below the critical value that secures diffusion equilibrium and, in susceptible individuals, which may even lead to pulmonary oedema and, therefore, an extended diffusion path length; and
• increased regional inequalities in the matching of alveolar ventilation to pulmonary capillary perfusion ($\dot{V}_A/\dot{Q}$).

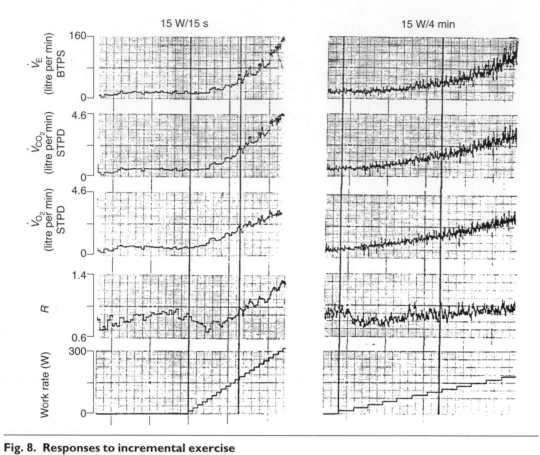

Fig. 8. Responses to incremental exercise

Ventilatory ($\dot{V}_E$) and pulmonary gas exchange ($\dot{V}_{O_2}$, $\dot{V}_{CO_2}$, R) responses to a rapid incremental exercise test (15 W per 15 s; left panel) and a slow incremental exercise test (15 W per 4 min; right panel) performed to the limit of tolerance on a cycle ergometer. The left-hand vertical line indicates the start of the test, and the right-hand vertical line indicates the lactate threshold (θ_L). For the rapid test, note the acceleration in the $\dot{V}_{CO_2}$ response at high work rates and the more prominent increase in R; this behaviour forms the basis for the non-invasive estimation of θ_L. Abbreviations used: BTPS, body temperature and pressure, saturated; SPTD, standard temperature and pressure, dry.

Cardiovascular transfer

The convective transfer of O_2 from the lungs to the working muscles is given by the product of $\dot{Q}_M$ and the arterial O_2 content (C_{aO_2}), and retains a close correlation with $\mu\dot{V}_{O_2}$. This has traditionally been regarded as being a major limiting factor to $\mu\dot{V}_{O_2}$.

The influence of C_{aO_2} on vascular O_2 transport is normally relatively minor, as P_{aO_2} typically does not fall acutely during exercise except, as discussed earlier, at high work rates in some very fit individuals. However, despite P_{aO_2} being located on the upper flat portion of the

O_2 dissociation curve (at sea level), small but significant increases in $\mu\dot{V}_{O_2}$ can occur with the breathing of 100% O_2 rather than room air (see [5] for further discussion). In addition, if haemoglobin levels are increased, $\mu\dot{V}_{O_2}$ also increases. This is the justification for performance-enhancement strategies, such as the re-infusion of autologous blood and the stimulation of red-cell production by administration of synthetic erythropoietin. Some investigators have also proposed an improved diffusive transfer of O_2 at the muscle because of reduced distances between adjacent red cells in capillar-

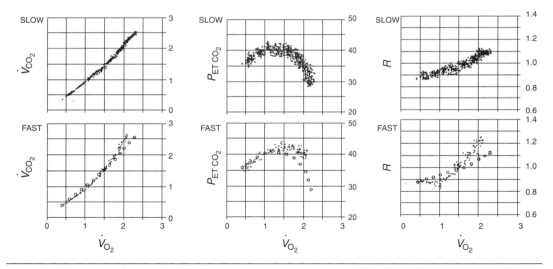

Fig. 9. Responses of $\dot{V}_{CO_2}$, end-tidal P_{CO_2} (P_{ETCO_2}) and R as a function of O_2 for 'slow' (upper panels) and 'fast' (lower panels) ramp protocols, for a single subject

For the purposes of comparison, the 'slow' response profiles (o; averaging interval, 0.25 litre per min $\dot{V}_{O_2}$) are also superimposed on the corresponding 'fast' protocol. Note the more marked evolution of CO_2 at high work rates in the 'fast' test (with both $\dot{V}_{CO_2}$ and R increasing more rapidly); however, respiratory compensation was not discernible (i.e. P_{ETCO_2} did not fall).

ies and, therefore, an increased capillary surface area for diffusion (eqn. 10). Conversely, challenges to vascular O_2 transport induced by conditions such as hypoxic hypoxia, stagnant hypoxia and anaemic hypoxia are all associated with not only a lowering of θ_L (as described earlier), but also of $\mu\dot{V}_{O_2}$.

As discussed in Chapter 7, the heart has been viewed for many years to be a major limiting factor to $\mu\dot{V}_{O_2}$ in humans. An important factor in this regard is heart size which, in turn, dictates stroke volume: larger stroke volumes, and therefore larger $\dot{Q}_T$ values, are associated with higher $\mu\dot{V}_{O_2}$ values. Cardiac output does, however, eventually evidence a maximum as progressively greater muscle mass is recruited for the activity (i.e. equivalent to 50% or so of the total muscle mass); as mentioned earlier, this is also the case for $\mu\dot{V}_{O_2}$.

It has been suggested that muscles have a greater potential to accommodate blood flow than is actually achieved at maximum exercise ([4,10]; see Chapter 7 in this volume). Indeed, this is one of the reasons put forward in support of the contention that the cardiac output is nor-

mally the most likely source of cardiovascular limitation to exercise. However, others dispute this view, based upon observations of $\dot{Q}_M$ during maximal contractions in isolated muscles compared with values during maximal volitional exercise.

There is also recent evidence that provides support for a vascular limitation to $\dot{V}_{O_2}$ kinetics during high-intensity exercise. For example, the slow $\dot{V}_{O_2}$ dynamics at these work rates can be speeded if the work is preceded by a bout of exercise sufficiently intense to induce metabolic acidaemia [21] (Fig. 7). This speeding is associated with a markedly reduced increment of both blood [lactate] and the degree of acidaemia during the exercise. Prior moderate exercise does not induce this effect. This effect has, therefore, been ascribed to a flow-dependence of $\dot{V}_{O_2}$ only above θ_L, as a result of improved muscle perfusion during the on-transient (consequent to the residual effects of vasodilator metabolites). An alternative explanation, however, is that this behaviour reflects an altered gain and/or τ of the high-energy-phosphate profiles at these work rates. What is uncertain at present is whether this speeding phenomenon also improves exercise tolerance — i.e. whether a given high-intensity work rate may be sustained for longer and whether $\mu\dot{V}_{O_2}$ may be increased.

Muscle transfer

Does the diffusive transfer of O_2 from the interior of the red blood cell in the skeletal muscle capillary bed to its intramitochondrial sites of utilization impose an additional level of constraint or limitation at these high work rates? This depends critically on the ability to estimate validly the individual elements in the Fick formulation:

$$\dot{V}_{O_2} = D_{O_2} \cdot A/l \cdot (P_{\bar{c}O_2} - P_{MO_2}) \qquad (10)$$

where D_{O_2} is the 'diffusion coefficient' for O_2, A is the surface area across which O_2 exchange takes place, l is the average path length between the capillary bed and the mitochondrial interior, $P_{\bar{c}O_2}$ is the mean P_{O_2} in the muscle capillary bed, and P_{MO_2} is the mean intramitochondrial P_{O_2} — which is assumed to be close or equal to zero at high work rates. As work rate increases, it is generally agreed that A increases and l decreases, as a result of a progressive recruitment of capillary bed. The behaviour of $P_{\bar{c}O_2}$ is less clear, however. Wagner [5] has stated that "muscle venous P_{O_2} and mean capillary P_{O_2} rise and fall together (as conditions are altered) and proportionally". This assumption can be challenged on two fundamental counts. First, it is difficult to envisage there being a constant proportionality between $P_{\bar{c}O_2}$ and P_{vO_2} even for a single capillary over a wide range of work rates. Factors such as the capillary transit time, the range of the sigmoid O_2 dissociation curve over which $P_{\bar{c}O_2}$ changes during the vascular transit and the magnitude of the Bohr effect are likely to change appreciably. These complexities are exacerbated by the heterogeneities in the regional distribution of perfusion relative to local metabolic rate in skeletal muscle. As a result, therefore, the extent to which this diffusive step in the 'O_2 cascade' is limiting at high work rates may not be simply resolved at present.

▶ The V_{O_2}–work rate relationship above θ_L therefore departs from its sub-θ_L linearity, being steeper between θ_L and θ_F and plateauing at the maximum V_{O_2} above θ_F.
▶ The mechanisms of the 'excess' V_{O2} are controversial, but appear to involve intramuscular mechanisms.
▶ Arterial hypoxaemia is seen in some highly fit individuals, because of diffusion impairment across the alveolar–capillary membrane resulting from critical shortening of pulmonary transit time.
▶ Limitations to convective O_2 transfer are likely to involve both the heart and the vascular bed in working muscle.
▶ The extent to which $\dot{Q}_{MO_2}$ is limited by intramuscular diffusion is unclear, because of uncertainties about the influence of regional differences in the distribution of muscle perfusion and metabolic rate.

CO_2 exchange

The factors determining the maximum rate of CO_2 output ($\mu\dot{V}_{CO_2}$) during high-intensity exercise have received far less attention. There is a growing recognition, however, that when the demand for pulmonary CO_2 clearance reaches high levels, the level of ventilation necessary to effect this clearance adequately (i.e. avoiding CO_2 retention) may approach or even exceed the maximal capacity of the respiratory system. This issue is taken up at length in Chapter 9 in this volume and also by Rowell [4] and Wagner [5]. It is appropriate here, however, to consider briefly the factors which determine the requirement for pulmonary CO_2 exchange and which, therefore, may contribute to limitation at high work rates (see [32,33] for discussion).

The primary sources of CO_2 generation above θ_L are: aerobic metabolism, the buffering of lactic acid by bicarbonate and the degree of respiratory compensation for the metabolic acidaemia (see [32,33,55] for further discussion). The latter two mechanisms are considered briefly below.

Buffering

An immediate consequence of metabolic acidosis is that the metabolic production of CO_2 in the working muscles is supplemented by additional CO_2 generated from the buffering of lactic acid by the bicarbonate system in muscle (mainly as potassium bicarbonate) and blood (chiefly as sodium bicarbonate):

$$CH_3CHOHCOO^-H^+ + Na^+HCO_3^-$$

Lactic acid   Sodium bicarbonate

$$\rightarrow CH_3CHOHCOO^-Na^+ + H_2CO_3 \quad (11)$$

Sodium lactate   Carbonic acid

The increase in arterial blood lactate concentration is essentially mirrored by a decrease in the arterial bicarbonate concentration ($[HCO_3^-]_a$). And, although Wasserman [1] has argued that a small early contribution to H^+ buffering derives from non-bicarbonate buffers, such as proteins and phosphate buffers (~0.5–1.0 mEq per litre), some 90% of the lactic acid formed at these work rates is buffered by the HCO_3^- system. Molecular CO_2 is evolved from the rapid dissociation of carbonic acid into CO_2 and H_2O. This CO_2, therefore, supplements the CO_2 produced from aerobic tissue metabolism.

As $\dot{V}_{CO_2}$ is now greater than $\dot{Q}_{CO_2}$, R exceeds the respiratory quotient. That is, $\dot{V}_{CO_2}$ increases faster than $\dot{V}_{O_2}$ with increasing work rate (Fig. 8 and Fig. 9). It is important to recognize that this release of CO_2 from the body CO_2 stores from this mechanism takes place only while the bicarbonate levels are falling. That is, if work rate increments are imposed for sufficiently long to allow $[HCO_3^-]_a$ to stabilize at each work rate, there will be no further release of CO_2 from the stores (this will have taken place in the early phase of the increment). Consequently, $\dot{V}_{CO_2}$ will once more equal $\dot{Q}_{O_2}$, and R will again equal the respiratory quotient; i.e. despite $[HCO_3^-]$ being reduced.

Respiratory compensation

Naturally, this 'buffering' of lactic acid does not actually prevent arterial pH from falling. An additional $\dot{V}_E$ drive is required to effect a compensatory reduction in P_{aCO_2} and, hence, constrains the fall of arterial and muscle pH (see Chapter 9 in this volume).

The higher the work rate, the greater is the contribution to the pulmonary CO_2 clearance requirement from each of these three sources. What may not be quite so obvious is that the clearance requirement at a particular high work rate also depends on the form of the exercise protocol. Rapidly incremented work rates result in an appreciably greater rate of CO_2 evolution from bicarbonate stores (see earlier; Fig. 8 and Fig. 9). However, the contribution from respiratory compensation is rather small, because of its relatively slow recruitment (Fig. 9). Therefore, P_{aCO_2} tends to increase, rather than fall. In contrast, for prolonged constant-load exercise or slow-incremental exercise, the buffering mechanisms are largely complete, while the respiratory compensation has had time to develop; hypocapnia is, therefore, evident. The removal of CO_2 from the arterial stores predominates under these conditions.

Given the influence of ramp incrementation rate on the response profile of P_{CO_2} above θ_L, the assertion that the slope of the supra-θ_L $\dot{V}_{CO_2}$–$\dot{V}_{O_2}$ relationship (ie. 'S2' of the 'V-slope' display) provides a useful index of the decrease in $[HCO_3^-]_a$ that occurs above θ_L — and, in turn, of the increase in blood [lactate] — should be subject to careful scrutiny [55]. This assertion is justified only if P_{aCO_2} can be assumed (or, better, demonstrated) to remain unchanged throughout the entire supra-θ_L domain. It is clear that this condition becomes increasingly less likely as the ramp incrementation rate is reduced.

The issue of whether there are limitations to CO_2 exchange at high work rates has received far less attention than for O_2 exchange. Under what conditions, for example, might $\dot{V}_{CO_2}$ attain a maximum? This question cannot easily be answered. The potentially limiting levels of pulmonary ventilation required by the high demands for CO_2 clearance in elite endurance athletes present a functional dilemma: should $\dot{V}_E$ be 'allowed' to encroach on its limit to maximize $\dot{V}_{CO_2}$ and limit the fall of pH in blood and muscle, but as a consequence exacerbate dyspnoeic sensations? Or, should the ventilatory increase be 'held in check' by constraining the magnitude of the respiratory compensation, thus attentuating the degree of dyspnoeic sensation but, therefore, 'allowing' greater acidosis and arterial hypoxaemia — and hence earlier limb-muscle fatigue? These issues remain to be resolved.

▶ The pulmonary CO_2 output ($\dot{V}_{CO_2}$) response during moderate exercise is dictated by muscle CO_2 production and the body's CO_2 storage capacity.

▶ Above θ_L, $\dot{V}_{CO_2}$ is supplemented by CO_2 from bicarbonate-buffering of lactic acid, and by respiratory compensation for the low arterial pH.

▶ CO_2 clearance may be impaired in highly fit individuals whose ventilatory requirement at high work rates may encroach upon (or even exceed) the mechanical limits of the lungs and chest wall.

Conclusions

▶ Interpretation of pulmonary gas exchange responses during dynamic exercise should take account of both the temporal and intensity domain of interest.

▶ The temporal profile of pulmonary O_2 and CO_2 exchange during exercise is determined by muscle metabolic rates for O_2 and CO_2, through the transform of intervening vascular transit delays and gas storage capacitances.

▶ In general, cardiovascular and intramuscular mechanisms impose limits on pulmonary O_2 exchange although, in some highly fit individuals, the lungs can become involved.

▶ Pulmonary CO_2 transfer is limited by the ventilatory control system and, under conditions in which ventilation approaches the system maximum, the associated symptoms of breathlessness, or dyspnoea.

Further reading

Barclay, J.K. (ed.) (1995) Symposium: Mechanisms which control $\dot{V}_{O_2}$ near $\dot{V}_{O_2max}$. *Med. Sci. Sports Exercise* **27**, 35–64

Gadian, D.G. (1982) *Nuclear Magnetic Resonance and its Application to Living Systems*, Clarendon Press, Oxford

Poole, D.C., Whipp, B.J. and Barstow, T.J. (eds.) (1994) Symposium: Mechanistic basis of the slow component of $\dot{V}_{O_2}$ kinetics during heavy exercise. *Med. Sci. Sports Exercise* **26**, 1319–1358

Rowell, L.B. and Shepherd, J.T. (eds.) (1996) *Handbook of Physiology, Section 12, Exercise: Regulation and Integration of Multiple Systems,* American Physiological Society, Bethesda

Steinacker, J.W. and Ward, S.A. (eds.) (1996) *The Physiology and Pathophysiology of Exercise Tolerance*, Plenum Press, New York

References

1. Wasserman, K. (1994) Coupling of external to cellular respiration during exercise: the wisdom of the body revisited. *Am. J. Physiol.* **266**, E519–E539

2. Saltin, B. and Gollnick, P.D. (1983) Skeletal muscle adaptability: significance for metabolism and performance. In *Handbook of Physiology: Skeletal Muscle* (Peachey, L.D., Adrian, R.H. and Geiger, S.R., eds.), pp. 555–631, American Physiological Society, Washington

3. Åstrand, P.-O. and Rodahl, K. (1977) *Textbook of Work Physiology*, McGraw-Hill, New York

4. Rowell, L.B. (1993) *Human Cardiovascular Control*, Oxford University Press, New York

5. Wagner, P.D. (1996) Determinants of maximal oxygen transport and utilization. *Annu. Rev. Physiol.* **58**, 21–50

6. Johnson, R.L., Jr., Heigenhauser, G.F., Hsia, C.C.W., Jones, N.L. and Wagner, P.D. (1996) Determinants of gas exchange and acid–base balance during exercise. In *Handbook of Physiology, Section 12, Exercise: Regulation and Integration of Multiple Systems* (Rowell, L.B. and Shepherd, J.T., eds.), pp. 515–584, American Physiological Society, Bethesda

7. Weibel, E.R. (1984) *The Pathway for Oxygen: Structure and Function in the Mammalian Respiratory System*, Harvard University Press, Cambridge

8. Shephard, R.J. (1977) *Endurance Fitness*, University of Toronto Press, Toronto

9. Laughlin, M.H., Korthuis, R.J., Duncker, D.J. and Bache, R.J. (1996) Control of blood flow to cardiac and skeletal muscle during exercise. In *Handbook of Physiology, Section 12, Exercise: Regulation and Integration of Multiple Systems* (Rowell, L.B. and Shepherd, J.T., eds.), pp. 705–769, American Physiological Society, Bethesda

10. Andersen, P. and Saltin, B. (1985) Maximal perfusion of skeletal muscle in man. *J. Physiol. (London)* **366**, 233–249

11. Molé, P.A., Coulson, R.L., Canton, J.R., Nichols, B.G. and Barstow, T.J. (1985) *In vivo* ^{31}P-NMR in human muscle: transient patterns with exercise. *J. Appl. Physiol.* **59**, 101–104

12. Yoshida, T. and Watari, H. (1993) ^{31}P-Nuclear magnetic resonance spectroscopy study of the time course of energy metabolism during exercise and recovery. *Eur. J. Appl. Physiol.* **66**, 494–499

13. Barstow, T.J., Buchthal, S., Zanconato, S. and Cooper, D.M. (1994) Muscle energetics and pulmonary oxygen uptake kinetics during moderate exercise. *J. Appl. Physiol.* **77**, 1742–1749

14. Rossiter, H.B., Ward, S.A., Doyle, V.L., Howe, F.A., Griffiths, J.R., and Whipp, B.J. (1999) Inferences from O_2 uptake with respect to intramuscular [PCr] kinetics during moderate exercise in humans. *J. Physiol.* **518**, 921–932

15. Mahler, M. (1985) First-order kinetics of muscle oxygen consumption, and an equivalent proportionality between $\dot{Q}_{O_2}$ and phosphorylcreatine level. *J. Gen. Physiol.* **86**, 135–165

16. Chance, B., Leigh, Jr, J.S., Clark, B.J., Maris, J., Kent, J., Nioka, S. and Smith, D. (1985) Control of oxidative metabolism and oxygen delivery in human skeletal muscle: a steady-state analysis of the work/energy cost transfer function. *Proc. Natl. Acad. Sci. U.S.A.* **82**, 8384–8388

17. Kushmerick, M.J., Meyer, R.A. and Brown, T.R. (1992) Regulation of oxygen consumption in fast- and slow-twitch muscle. *Am. J. Physiol.* **263**, C598–C606

18. Meyer, R.A. and Foley, J.M. (1996) Cellular processes integrating the metabolic response to exercise. In *Handbook of Physiology, Section 12, Exercise: Regulation and Integration of Multiple Systems* (Rowell, L.B. and Shepherd, J.T., eds.), pp. 841–869, American Physiological Society, Bethesda

19. Hughson, R.L., Green, H.J., Phillips, S.M. and Shoemaker, J.K. (1996) Physiological limitations to endurance exercise. In *The Physiology and Pathophysiology of Exercise Tolerance* (Steinacker, J.W. and Ward, S.A., eds.), pp. 211–217, Plenum Press, New York

20. Williamson, J.W., Raven, P.B., Foresman, B.H. and Whipp, B.J. (1993) Evidence for an intramuscular ventilatory stimulus during dynamic exercise in man. *Respir. Physiol.* **94**, 121–135

21. Gerbino, A., Ward, S.A., and Whipp, B.J. (1996) Effects of prior exercise on pulmonary gas exchange kinetics during high-intensity exercise in humans. *J. Appl. Physiol.* **80**, 99–107

22. Yoshida, T. and Whipp, B.J. (1994) Dynamic asymmetries of cardiac output transients in response to muscular exercise in man. *J. Physiol. (London)* **480**, 355–359

23. Whipp, B.J., Ward, S.A., Smith, R.E. and Hussain, S.T. (1995) The dynamics of pulmonary O_2 uptake and femoral artery blood flow during moderate intensity exercise in humans. *J. Physiol. (London)* **483P**, 130P

24. Whipp, B.J. and Ward, S.A. (1990) Physiological determinants of pulmonary gas exchange kinetics during exercise. *Med. Sci. Sports Exercise* **22**, 62–71

25. Whipp, B.J. (1994) The slow component of O_2 uptake kinetics during heavy exercise. *Med. Sci. Sports Ex.* **26**, 1319–1326

26. Casaburi, R., Daly, J., Hansen, J.E. and Effros, R.M. (1989) Abrupt changes in mixed venous blood gas composition following onset of exercise. *J. Appl. Physiol.* **67**, 1106–1112

27. Cerretelli, P. and Di Prampero, P.E. (1987) Gas exchange in exercise. In *Handbook of Physiology. The Respiratory System*, section 3, vol. IV (Farhi, L.E. and Tenney, S.M., eds.), pp. 297–339, American Physiological Society, Bethesda

28. Linnarsson, D. (1974) Dynamics of pulmonary gas exchange and heart rate at start and end of exercise. *Acta Physiol. Scand.* **415** (suppl.), 1–68

29. Paterson, D.H. and Whipp, B.J. (1991) Asymmetries of oxygen uptake transients at the on- and off-set of heavy exercise in humans. *J. Physiol. (London)* **443**, 575–586

30. Hagberg, J.M., Hickson, R.C., Ehsani, A.A. and Holloszy, J.O. (1980) Faster adjustment to and from recovery from submaximal exercise in the trained state. *J. Appl. Physiol.* **48**, 218–22

31. Babcock, M.A., Paterson, D.H., Cunningham, D.A. and Dickinson, J.R. (1994) Exercise on-transient gas exchange kinetics are slowed as a function of age. *Med. Sci. Sports Exercise* **26**, 440–446

32. Whipp, B.J. (1987) Dynamics of pulmonary gas exchange during exercise in man. *Circulation* **VI**, 18–28

33. Wasserman, K., Hansen, J.E., Sue, D.Y., Whipp, B.J. and Casaburi, R (1994) *Principles of Exercise Testing and Interpretation*, Lea and Febiger, Philadelphia

34. Whipp, B.J., Davis, J.A., Torres, F. and Wasserman, K. (1981) A test to determine the parameters of aerobic function during exercise. *J. Appl. Physiol.* **50**, 217–221

35. Barstow, T.J., Lamarra, N. and Whipp, B.J. (1990) Modulation of muscle and pulmonary O_2 uptakes by circulatory dynamics during exercise. *J. Appl. Physiol.* **68**, 979–989

36. Grassi, B., Poole, D.C., Richardson, R.S., Knight, D.R., Kipp Erickson, B. and Wagner, P.D. (1996) Muscle O_2 kinetics in humans: implications for metabolic control. *J. Appl. Physiol.* **80**, 988–998

37. McCreary, C.R., Chilibeck, P.D., Marsh, G.D., Paterson, D.H., Cunningham, D.A. and Thompson, R.T. (1996) Kinetics of pulmonary oxygen uptake and muscle phosphates during moderate-intensity calf exercise. *J. Appl. Physiol.* **81**, 1331–1338

38. Hughson, R.L., Weisiger, K.H. and Swanson, G.D. (1987) Blood lactate concentration increases as a continuous function in progressive exercise. *J. Appl. Physiol.* **62**, 1975–1981

39. Dennis, S.C., Noakes, T.D. and Bosch, A.P. (1992) Ventilation and blood lactate increase exponentially during incremental exercise. *J. Sports Sci.* **10**, 437–449

40. Brooks, G.A. (1985) Anaerobic threshold: review of the concept and directions for future research. *Med. Sci. Sports Exercise* **17**, 22–31

41. Davis, J.A. (1985) Anaerobic threshold: review of the concept and directions for future research. *Med. Sci. Sports Exercise* **17**, 6–18

42. Whipp, B.J. and Wasserman, K. (1994). Exercise. In *Textbook of Respiratory Medicine* (Murray, J.F. and Nadel, J.A., eds.), pp. 219–250, W.B. Saunders, Philadelphia

43. Gladden, L.B. (1996) Lactate transport and exchange during exercise. In *Handbook of Physiology. Section 12, Exercise: Regulation and Integration of Multiple Systems* (Rowell, L.B. and Shepherd, J.T., eds.), pp. 614–648, American Physiological Society, Bethesda

44. Barstow, T.J. and Mol, P.A. (1991) Linear and nonlinear characteristics of oxygen uptake kinetics during heavy exercise. *J. Appl. Physiol.* **71**, 2199–2106

45. Whipp, B.J. and Mahler, M. (1980) Dynamics of pulmonary gas exchange during exercise. In *Pulmonary Gas Exchange* (West, J.B., ed), pp. 33–96, Academic Press, New York

46. Barstow, T.J. (1994) Characterization of $\dot{V}_{O_2}$ kinetics during heavy exercise. *Med. Sci. Sports Exercise* **26**, 1327–1334

47. Langsetmo, I., Weigle, G.E., Fedde, M.R., Erickson, H.H., Barstow, T.J. and Poole, D.C. (1997) $\dot{V}_{O_2}$ kinetics in the

horse during moderate and heavy exercise. *J. Appl. Physiol.* **83**, 1235–1241

48. Poole, D.C., Ward, S.A., Gardner, G.W. and Whipp, B.J. (1988) Metabolic and respiratory profile of the upper limit for prolonged exercise in man. *Ergonomics* **31**, 1265–1279

49. Medbø, J.I., Mohn, A., Tabata, I., Bahr, R. and Sejersted, O. (1998) Anaerobic capacity determined by the maximal accumulated oxygen deficit. *J. Appl. Physiol.* **64**, 50–60

50. Binzone, T., Ferretti, G., Schenker, K. and Cerretelli, P. (1992) Phosphocreatine hydrolysis by ^{31}P-NMR at the onset of constant-load exercise in humans. *J. Appl. Physiol.* **73**, 1644–1649

51. Ward, S.A., Groom, R. and Whipp, B.J. (1995) Kinetics of skeletal muscle oxygenation and oxygen uptake during exercise in humans. *J. Sports Sci.* **14**, 105

52. Belardinelli, R., Barstow, T.J., Porszasz, J. and Wasserman, K. (1995) Skeletal muscle oxygenation during constant work rate exercise. *Med. Sci. Sports Exercise* **27**, 512–519

53. Poole, D.C., Schaffartzik, W., Knight, D.R., Derion, T., Kennedy, B., Guy, H.J., Prediletto, R. and Wagner, P.D. (1991) Contribution of exercising legs to the slow component of oxygen uptake kinetics in humans. *J. Appl. Physiol.* **71**, 1245–1253

54. Barstow, T.J., Jones, A.M., Nguyen, P.H. and Casaburi, R. (1996) Influence of muscle fiber type and pedal frequency on oxygen uptake kinetics of heavy exercise. *J. Appl. Physiol.* **81**, 1642–1650

55. Ward, S.A. and Whipp, B.J. (1992) Influence of body CO_2 stores on ventilatory–metabolic coupling during exercise. In *Control of Breathing and Its Modeling Perspective* (Honda, Y., Miyamoto, Y., Konno, K. and Widdicombe, J.G., eds.), pp. 425–431, Plenum Press, New York

9

The performance of the pulmonary system during exercise in athletes

Brian J. Whipp

Department of Physiology, St George's Hospital Medical School, Cranmer Terrace, London SW17 0RE, U.K.

Introduction

It is conventional to consider the primary function of the lung to be that of 'arterializing' the mixed venous blood. However, a better definition might be that its function is to 'alveolarize' the blood: the closeness of the arterial blood partial pressure to that of the alveolar gas then becomes a quantifiable index of how well the lung has subserved its function.

We shall consider in this chapter how well this gas-exchange function of the lung is achieved in athletic subjects, i.e. subjects who have the ability to attain high levels of metabolic rate with the consequent demands for high levels of pulmonary fluid flow (air and blood). As frames of reference we shall consider (i) the requirement for blood-gas and acid–base regulation; (ii) the cost of meeting these requirements; and (iii) the extent to which the pulmonary system's response is constrained or limited during maximum exercise.

Ventilatory requirements of exercise

To maintain alveolar, and hence arterial, blood–gas partial pressures at or close to resting levels, ventilation must increase in proportion to metabolic rate as shown schematically in Fig. 1. With respect to CO_2 exchange:

$$F_{A CO_2} = \frac{\dot{V}_{CO_2} (STPD)}{\dot{V}_A} \quad (STPD) \qquad (1)$$

where $F_{A CO_2}$ is the fractional concentration of alveolar gas and V is the flow or volume per unit time. Note that in this equation both gas volumes are determined under the same conditions, i.e. standard temperature (0°C) and pressure (1 atm); dry (STPD).

Convention has, reasonably, dictated that ventilatory volumes are expressed under the conditions that reflect how much air was actually moved, i.e. at body temperature and pressure, saturated with water vapour (BTPS). And as the partial pressure (P) of the gas is of more interest physiologically, then:

$$P_{A CO_2} = \frac{863 \times \dot{V}_{CO_2} (STPD)}{\dot{V}_A \quad (BTPS)} \qquad (2)$$

Similarly, for oxygen exchange:

$$P_{A O_2} = P_{i O_2} - \frac{863 \times \dot{V}_{O_2} (STPD)}{\dot{V}_A \quad (BTPS)} \qquad (3)$$

where P_{iO_2} is the P_{O_2} of inspired air, and 863 is the constant which corrects for: (a) the different conditions of reporting the gas volumes and (b) the transformation of the fractional concentration to the partial pressure.

While the concept of the oxygen uptake ($\dot{V}_{O_2}$) and the CO_2 output ($\dot{V}_{CO_2}$) are relatively straightforward, neither $\dot{V}_A$ nor the alveolar gas partial pressures in eqns. 2 and 3 are quite as simple. As may be imagined in a structure as complex as the lung, which has significant regional variations of alveolar P_{O_2} and P_{CO_2} ($P_{A O_2}$ and $P_{A CO_2}$, respectively), it is difficult to establish a single average value for alveolar gas tensions. Similarly, the problem of

135

regional differences in alveolar ventilation ($\dot{V}_A$) to perfusion ($\dot{Q}$), especially in subjects who have a component of alveolar deadspace (i.e. local $\dot{V}_A/\dot{Q} = \infty$), complicates the determination of $\dot{V}_A$. This difficulty is conventionally overcome by the expedient of assuming that P_{ACO_2} is exactly equal to the arterial P_{CO_2} (P_{aCO_2}). This allows alveolar ventilation to be computed readily from eqn 2. However, it is important to realize that this alveolar ventilation is, in a sense, figmentary; it may not be the actual level of ventilation to the alveoli in a subject, rather it is the alveolar ventilation that would establish a P_{ACO_2} exactly equal to the current level of P_{aCO_2}. Similarly, in eqn 3, the P_{AO_2} that is computed is that of the ideal lung, in which there would be no difference between P_{AO_2} and P_{aO_2}, i.e. as for the P_{CO_2} calculation. In reality the 'real' P_{AO_2} is likely to be higher than the 'ideal' alveolar value calculated from eqn 3.

However, from eqns. 2 and 3 it is clear that P_{ACO_2} and P_{AO_2} can be maintained at a constant level during exercise only if V_A changes in precise proportion to $\dot{V}_{CO_2}$ and $\dot{V}_{O_2}$ respectively. Note, however, that V_A is common to both equations, i.e.

$$\frac{863 \times \dot{V}_{CO_2-}}{P_{ACO_2}} <= \dot{V}_A => \frac{863 \times \dot{V}_{O_2-}}{P_{iO_2} - P_{AO_2}} \quad (4)$$

In this equation, we have neglected the effect on P_{O_2} of the slight difference in inspiratory and expiratory levels of ventilation that occur when $R \neq 1$, since the effect is relatively small [1] and does not materially affect the argument.

Under conditions in which the CO_2 exchange rate differs from that of O_2, either because of differences in substrate-utilization profiles [2] or because of transient variations in the body gas stores (Fig. 2), alveolar ventilation obviously cannot meet the demands of both O_2 and CO_2 exchange, as shown in Fig. 1 and eqn 4. It has been widely demonstrated that, under such conditions, ventilation changes in closer proportion to the CO_2 exchange than to oxygen uptake [2,3]. Consequently, P_{ACO_2} is the more closely regulated variable, with P_{AO_2} being allowed to change as a consequence. But as these P_{O_2} changes normally only vary over a

range in which the oxyhaemoglobin dissociation curve is relatively flat, then changes in O_2 content or saturation of the arterial blood will not be affected to any great extent by these changes.

It is, therefore, appropriate to consider the demands for ventilation during exercise in an athlete using CO_2 exchange as the frame of reference. From the left segment of eqn 4, it may be seen that the demands for alveolar ventilation increase as a linear function of $\dot{V}_{CO_2}$ — at any set-point level of P_{ACO_2}. Consequently, the greater the CO_2 output, the greater is the requirement for ventilation. However, when P_{CO_2} is either regulated at a lower level or is lowered to provide compensation for a metabolic acidaemia during exercise, then the alveolar ventilation must be appropriately higher — for any given level of CO_2 output. The demands for ventilation, however, are not to provide alveolar ventilation but total pulmonary ventilation. That is, it is also necessary to ventilate the deadspace of the lung. Therefore, eqn 4 must be modified to take account of this effect.

As alveolar ventilation is simply the difference between the total ventilation of the lung ($\dot{V}_E$) and the ventilation of the deadspace (V_D), the alveolar ventilation may be written as:

$$\dot{V}_A = \dot{V}_E - \dot{V}_D \quad (5)$$

or

$$\dot{V}_A = \dot{V}_E (1 - V_D/V_T) \quad (6)$$

where V_D/V_T is the physiological deadspace fraction of the breath.

The mass balance equation, therefore, linking ventilation to pulmonary CO_2 exchange is:

$$\dot{V}_E = \frac{863 \times \dot{V}_{CO_2}}{P_{ACO_2}(1 - V_D/V_T)} \quad (7)$$

Ventilation during muscular exercise is, therefore, determined by an interaction among the

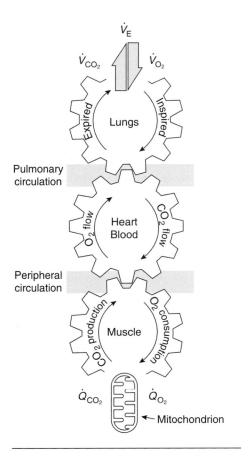

Fig. I. Schematic represeantation (using Wasserman's gears analogy) of the sytems linking metabolic O_2 consumption ($\dot{Q}_{O_2}$) and CO_2 production ($\dot{Q}_{CO_2}$) with pulmonary O_2 uptake ($\dot{V}_{O_2}$) and CO_2 output ($\dot{V}_{CO_2}$)

Note that pulmonary ventilation ($\dot{V}_E$) must sub-serve the demands of both O_2 and CO_2 exchange.

three defining variables, $\dot{V}_{CO_2}$, $P_{A\,CO_2}$ and V_D/V_T. The influence of variations in these determinants on ventilation is shown in Fig. 3.

Only in the steady state of muscular exercise does $\dot{V}_{CO_2}$ measured at the lung equal the CO_2 production rate in the tissues ($\dot{Q}_{CO_2}$); under these conditions, the respiratory gas exchange ratio (R) equals the metabolic respiratory quotient (RQ). Under non-steady-state conditions, however, the transient changes in the body CO_2 stores dissociate $\dot{V}_{CO_2}$ from $\dot{Q}_{CO_2}$. For example, during the on-transient of constant-load exercise, some of the metabolically produced CO_2 never reaches the lung for

exchange. This is because of the capacitative storage of CO_2 in the muscle. The $\dot{V}_{CO_2}$ measured at the lungs is, therefore, less than $\dot{Q}_{CO_2}$ during this phase. And, since changes in the muscle O_2 stores are trivially small with respect to those of CO_2, the gas exchange ratio R falls transiently [1,2], reaching a minimum at the point of the maximum rate of CO_2 storage. It subsequently rises again to equal the new metabolic steady-state RQ as the muscle P_{CO_2} stabilizes at its new, and higher, exercise value [1,2].

The opposite occurs at the off-transient. The CO_2 stores now discharge; this leads to the pulmonary R increasing to levels above that of the metabolic RQ. The ventilatory changes during these transients closely match those of the $\dot{V}_{CO_2}$. Consequently, during the phase in which R falls, the ventilatory increase is not appropriate for the $\dot{V}_{O_2}$; alveolar and arterial P_{O_2} are consequently reduced. This largely accounts for the transient hypoxaemia that has been described during the non-steady-state, on-transient phase of constant-load exercise (see [4] for discussion).

At work rates above the lactate threshold (θ_L), pulmonary CO_2 exchange is increased further [5], from two additional sources. First, CO_2 is produced as a result of the bicarbonate handling the proton, which is formed in concert with the increased lactate at these work rates, i.e.

$$CH_3 \cdot CHOH \cdot COO^- + H^+ + NaHCO_3$$
$$CH_3 \cdot CHOH \cdot COONa + H_2CO_3$$
$$\swarrow \searrow$$
$$HO + CO_2 \quad (8)$$

It is important to recognize here that the extra CO_2 that is formed in these reactions is quantitatively large. This may be readily discerned from the recognition that the glycogen-utilization rate must increase more than 12-fold under 'anaerobic' conditions to sustain a given ATP-production rate. The complete aerobic catabolism of one glucosyl unit of glycogen to 6 CO_2 and 6 H_2O yields 37 ATP molecules; its breakdown to lactate and its associated proton yields only 3 ATP molecules. Glycolytic flux must,

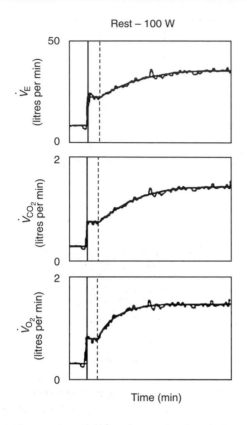

Rest – 100 W

Fig. 2. Time course of the responses of ventilation ($\dot{V}_E$), carbon dioxide output ($\dot{V}_{CO_2}$) and oxygen uptake ($\dot{V}_{O_2}$) in response to a constant work rate of 100 W from rest on a cycle ergometer

The solid vertical line reflects the onset of the exercise, and the dashed vertical line represents the phase-I–phase-II transition. Note that during phase-II, CO_2 output changes appreciably more slowly than does O_2 uptake as a consequence of the high tissue CO_2 capacitance. The dynamics of ventilation are highly correlated with those of CO_2 output during this phase.

therefore, increase by 12.3-fold (i.e. $37 \div 3$) to sustain the ATP-production rate. The proton production accompanying the resulting 24.6 milli-equivalent (mEq) production of lactate will, therefore, decrease the HCO_3^- by ~22 mEq (i.e. HCO_3^- only accounts for some 90% of this buffering phosphate: protein and phosphate buffers account for the remainder). This yields ~22 mM of additional CO_2. Of this, however, 6 mM replaces the CO_2 that would

have been produced aerobically for this rate of ATP formation. The net increase in CO_2 production rate is therefore:

22 mM anaerobic CO_2 – 6 mM aerobic CO_2
= 16 mM net CO_2 yield

This represents a ~2.5-fold increase in $\dot{V}_{CO_2}$ (for these reactions) during supra-θ_L exercise.

The increase in the extra amount of CO_2 from these reactions is naturally a direct function of the amount of decrease in the bicarbonate in both the blood and the muscle compartments. That is, any buffering that occurs through non-bicarbonate mechanisms, such as phosphate and protein, while important for H^+ regulation, does not produce extra CO_2. However, it is important to recognize that the rate at which extra CO_2 is produced from these reactions will be proportional to the rate at which the bicarbonate falls:

$$V_{CO_2} \propto f(\Delta HCO_3^-);$$

whereas $\dot{V}_{CO_2} \propto f(\dot{HCO_3}^-)$ \hfill (9)

Consequently, the more rapid the rate of rise of lactate, the greater is the increase in CO_2 production rate ($\dot{V}_{CO_2}$). This accounts for both $\dot{V}_{CO_2}$ and R being appreciably higher above θ_L during maximum incremental exercise, when the incrementation rate is rapid, than when the increment rate is relatively slow — because the rate of lactate increase, and hence bicarbonate decrease, is greater.

The second determining variable in eqn 7 (P_{aCO_2}) is normally regulated at, or near, resting levels in the steady state of moderate-intensity exercise. Hyperventilation can occur transiently, however, at the onset of exercise, especially in 'excitable' subjects; this is more common in treadmill than in cycle ergometry. But as the time constant for $\dot{V}_E$ ($\tau\dot{V}_E$) during the subsequent increase to the steady-state is slightly longer than that of $\dot{V}_{CO_2}$ ($\tau\dot{V}_{CO_2}$), a small and transient increase in P_{aCO_2} is both predictable and has been measured. The ratio $\tau\dot{V}_E/\tau\dot{V}_{CO_2}$ (i.e. a determinate of the magnitude

$$\dot{V}_E = \frac{863 \times \dot{V}_{CO_2}}{P_{aCO_2}(1 - V_D/V_T)}$$

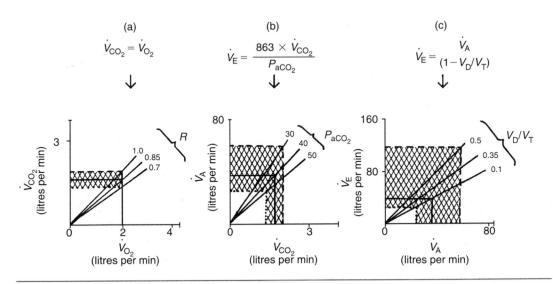

Fig. 3. Representation of the influence of variables on ventilation during muscular exercise

(a) A work rate requiring an O_2 uptake of 2 litres per min is considered. The CO_2 output for this work rate is determined by the respiratory gas exchange ratio, R. (b) The influence of the set-point for arterial P_{CO_2} on the alveolar–ventilatory response to this range of CO_2 outputs. (c) The influence of the dead-space fraction of the tidal volume on the ventilatory demands for the task is presented for the range of alveolar ventilations established in (b).

of the transient error in P_{aCO_2}) seems to depend, in large part, on the sensitivity of the subject's peripheral chemoreceptors, especially the carotid bodies [3]. When carotid body sensitivity is high (e.g. with experimentally induced hypoxia or metabolic acidaemia), $\tau\dot{V}_E$ is short; whereas, when carotid body sensitivity is low (e.g. with hyperoxia, pharmacological suppression with drugs such as dopamine, or when they have been surgically resected), $\tau\dot{V}_E$ is long.

Although P_{aCO_2} appears to be a regulated variable during moderate exercise, it must be lowered by hyperventilation to constrain the fall of pH at levels of exercise which induce the metabolic acidaemia. This compensatory decrease in P_{aCO_2} washes CO_2 out of the body stores and provides an additional source of extra CO_2 at high work rates, i.e.

$$pH = pK' + \frac{\log HCO_3^-}{\alpha \cdot P_{aCO_3^-}} \quad (10)$$

where α is the CO_2 solubility coefficient which relates P_{aCO_2} in mmHg to CO_2 content in mM per litre. An alternative way of considering this relationship is to replace P_{aCO_2} by its ventilatory and gas-exchange determinants, i.e.

$$pHa = pK' + \log\left\{\left[\frac{[HCO_3^-]}{25.8}_{(a)}\right] \times \left[\frac{\dot{V}_E}{\dot{V}_{CO_2}}\right] \times \left[(1-V_D/V_T)\right]\right\} \quad (11)$$

where the bracketed terms may be considered to represent the metabolic 'set point', the 'con-

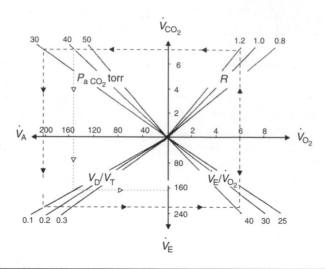

Fig. 4. Schematic representation of the influence of CO$_2$ output ($\dot{V}_{CO_2}$), arterial P_{CO_2} (P_{aCO_2}) and the dead-space fraction of the breath (V_D/V_T) on the ventilatory ($\dot{V}_E$) requirement for exercise

O$_2$ uptake ($\dot{V}_{O_2}$) and $\dot{V}_{CO_2}$ are measured in litre per min STPD and alveolar ventilation ($\dot{V}_A$) and minute ventilation ($\dot{V}_E$) are measured in litre per min BTPS. Note that the particular combination of these determining variables can alter substantially the $\dot{V}_E$ requirement at a given $\dot{V}_{O_2}$ (see text for further discussion).

trol' component and the ventilatory 'efficiency' terms, respectively.

The increase in $\dot{V}_{CO_2}$ is not a function of the magnitude of the decrease in P_{CO_2} caused by the hyperventilation, but rather results from the rate at which the P_{CO_2} falls, i.e. the rate at which CO$_2$ is evolved from the body CO$_2$ stores. However, only when all these defining variables (eqn 7) are considered together does one get an adequate sense of their integrated influence on establishing the ventilatory demands of muscular exercise. The effects at relatively low metabolic rate are presented in Fig. 3.

The combined influence of $\dot{V}_{CO_2}$, P_{aCO_2} and V_D/V_T on an athlete at a high level of $\dot{V}_{O_2}$ are presented in Fig. 4. This figure may seem daunting at first sight but it rewards the effort to work through it, quadrant by quadrant. In the upper-right quadrant, one begins by moving from the exercise oxygen uptake to its consequent CO$_2$ output; this is, naturally, related to the gas exchange ratio R. Once the $\dot{V}_{CO_2}$ for the particular work rate has been established, one might then ask what is the required alveolar ventilation for the task? As discussed earlier,

this depends upon the current level of P_{aCO_2}, which is presented in the top-left quadrant. At any given level of $\dot{V}_{CO_2}$, the alveolar ventilation required is a function of P_{aCO_2}: the lower the P_{CO_2}, the higher the alveolar ventilation. However, as it is the total ventilatory demand that is important— and not just that of alveolar ventilation — one moves to the bottom-left quadrant. This leads to the total ventilation required. This is a function of the current level of V_D/V_T: the higher the V_D/V_T, the greater is the ventilation required. It is quite instructive to begin at any level of O$_2$ and proceed to the ventilation required by considering reasonable combinations of R, P_{ACO_2} and V_D/V_T. This will make it quite apparent that the ventilatory requirement for a given level of exercise can vary enormously.

As the main focus of this chapter is the elite athlete, a high — but no longer implausibly high — $\dot{V}_{O_2}$ of 6 litre per min is considered. At maximum exercise, an R of 1.2 is quite reasonable; CO$_2$ output is consequently 7.2 litre per min. But as the subject is likely to be hyperventilating at such high levels of exercise, a reasonable P_{CO_2} of 30 mmHg has been chosen. Note that this demands an alveolar ventilation of over 200 litre per min. But V_D/V_T is likely to have fallen from its resting value of approximately 0.3–0.1 at high-intensity exercise.

The actual dead-space volume increases during exercise as a result of (i) the increased transmural pressure across the intra-pulmonary airways at end-inspiration (consequent to the more-negative intrapleural pressure causing airway distension), and (ii) some bronchodilatation that can occur as a result of the exercise-induced increase in catecholamines. These airway volume increases, however, are small compared with the size of the tidal volume increase; V_D/V_T decreases despite both V_D and V_T increasing. Taking V_D/V_T at maximum exercise to be 0.1, the consequent ventilatory demand of this work rate, under these conditions, will be 225 litres per min. Note also in Fig. 4 that, were the athlete not to hyperventilate but rather to maintain P_{aCO_2} at the approximate resting level of 40 mmHg, this would 'conserve' more than 50 litres per min of ventilation. The consequence in this case would be an appreciably greater fall in pH. For example, assuming arterial blood lactate to be 13 mEq per litre, with a consequent arterial bicarbonate level of 12 mEq per litre, arterial pH would be 7.22 in the case where the P_{aCO_2} were reduced to 30 mmHg, but would fall to 7.1 without this 10 mmHg hyperventilation. Consequently, the elite athlete is confronted with a dilemma during high-intensity exercise: providing a compensatory hyperventilation stresses ventilatory mechanics, whereas failure to hyperventilate leads to a more-precipitous fall of arterial and muscle pH.

It is pertinent to consider whether the high levels of ventilation required by the athlete in this example at maximum exercise cause the athlete to reach the mechanical limits of airflow generation. Consider in Fig. 4 that the subject's 225 litres per min ventilation, under these conditions, is apportioned into a tidal volume of 4.5 litres and a breathing frequency of 50 breaths per min; this requires an average breath duration of 1.2 s. Apportioning this total breath time into its typical inspiratory and expiratory components (0.7 s for inspiration, 0.5 s

for expiration) means that a 4.5 litre volume must be exhaled in 0.5 s. The mean expiratory flow is therefore some 9 litres per s. Since the expiratory airflow profile is roughly sinusoidal at these breathing frequencies, the peak expiratory airflow demanded under these conditions will be about 14 litres per s [i.e. $(\pi/2) \times 9$], which is enormously high. Does this demand for airflow, therefore, actually exceed the athlete's capacity to generate it? As will be discussed subsequently, this depends in large part on genetic structural aspects of the athlete's pulmonary system.

▶ The appropriateness of the ventilatory response to muscular exercise is best considered with respect to its effectiveness in regulating arterial blood gas tensions during moderate exercise.

▶ Under conditions in which the pulmonary gas-exchange rates of O_2 and CO_2 differ, as a result of either differences in substrate utilization or transient gas-storage differences, ventilation remains closely coupled to the CO_2 exchange rate. Consequently, arterial P_{CO_2} is more tightly regulated than is arterial P_{O_2}.

▶ During heavy exercise, associated with a metabolic acidosis, a further increment in ventilation is required to provide a compensatory decrease in arterial P_{CO_2} to constrain the fall in arterial pH.

The costs of meeting the ventilatory requirements

Ventilatory costs

Mechanical costs

The respiratory power (i.e. work rate) generated by the respiratory muscles during breathing is manifest as the pressure changes (P_{mus}) which distend the chest wall and the lungs and also produce airflow through the conducting airways (and a small tissue-flow component). There is also a small accelerative component (or inertance), which is often disregarded as it is,

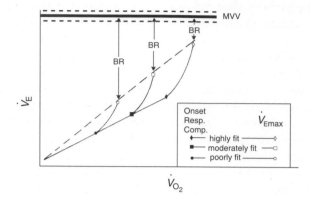

Fig. 5. Schematic representation of the $\dot{V}_E$ response to exhausting incremental exercise, for poorly fit (circles), normal (squares) and highly fit (diamonds) subjects

The $\dot{V}_E$ responses are similar below the lactate threshold (filled symbols), regardless of fitness. $\dot{V}_E$ at maximum exercise ($\dot{V}_{Emax}$), however, is progressively higher the greater the maximum $\dot{V}_{O_2}$ (open symbols): the 'breathing reserve' (BR) is therefore progressively reduced (BR = MVV − $\dot{V}_{Emax}$). The MVV is largely independent of fitness and training. The horizontal dashed lines above and below the MVV reflect that this value can be quite different, even in subjects of similar size. Modified with permission from [15]. ©1986, The American Physiological Society.

quantitatively, such a small component of the total, i.e. although the chest wall has a large mass, its acceleration during breathing is low, and although the acceleration of the inspired gas can be high, its mass is low.

The pressure required to overcome the total impedance of the pulmonary system is therefore:

$$P_{\text{mus}} \doteq E \cdot V + R \cdot \dot{V} + I \cdot \ddot{V} \qquad (12)$$

where E, R and I are the system elastance, resistance and inertance, and V, $\dot{V}$ and $\ddot{V}$ represent the volume, flow and acceleration terms respectively.

The respiratory power generated during exercise has been shown to increase curvilinearly with respect to increasing $\dot{V}_E$. Consequently, as the work rate increases progressively, greater increments in power are required to produce a given increase in $\dot{V}_E$ [6,7] However, in moderately fit subjects exercising at near-maximum levels, the respiratory muscle power is only ~30–40% of that achieved during

a maximal volitional ventilation (MVV) manoeuvre [8,9], i.e. the upper limits for the $\dot{V}_E$ that may be volitionally attained for periods as short as 15 s. Also, the $\dot{V}_E$ actually achieved during maximum exercise is appreciably less than the MVV in such subjects. This difference between maximum exercise $\dot{V}_E$ and the subject's MVV [10] has been termed the 'breathing reserve' (Fig. 5). Since the MVV is largely independent of fitness and training status, highly fit athletes (normalized for gender, age and height) typically have a lower breathing reserve than do the less-fit subjects.

The greater mechanical and metabolic costs of ventilation at high work rates are dependent in part on: (i) the increased contributions from turbulence (and even inertia) when air flow is high [11]; (ii) an increased elastic work of breathing, owing to a decreased compliance of the lungs, even in the tidal range, during exercise [12], an effect ascribed to the increase in pulmonary blood volume; (iii) the progressive increase of end-inspiratory lung volume that occurs as V_T increases during exercise tends to encroach on the upper, poorly compliant region of the compliance curve — although this effect will be ameliorated by the decrease in end-expiratory lung volume; (iv) the recruitment at high levels of ventilation of respiratory muscles with low mechanical efficiencies; and (v) the fact that chest wall 'distortion' [13] and the inertial component of the total impedance may no longer be negligible [14] at very high levels of ventilation.

Metabolic costs

The mechanical costs of the exercise hyperpnoea require both an increased respiratory muscle work rate and respiratory muscle O_2 consumption ($\dot{V}_{rmO_2}$). However, the relationship between $\dot{V}_{rmO_2}$ and $\dot{V}_E$ is not linear; it is concave upwards [15,16]. The estimation of $\dot{V}_{rmO_2}$ in humans, however, is technically quite difficult, as it represents such a small fraction of the whole-body $\dot{V}_{O_2}$.

A further difficulty results from the fact that if P_{aCO_2} is not maintained precisely, in tests which attempt to reproduce the exercise hyperpnoea with the subject at rest, then the whole-body $\dot{V}_{O_2}$ can change — not solely as a consequence of the increased respiratory muscle work rate but as a result of the altered pH. For example, with an induced respiratory alkalosis, whole-body V_{O_2} increases by ~10% per 10 mmHg reduction in P_{aCO_2} [17].

Although $\dot{V}_{rmO_2}$ is small compared with resting levels of V_E, it can be a significant component at high levels of exercise in highly fit athletes. Even in moderately fit subjects, a $\dot{V}_{rmO_2}$ of ~0.5 litre per min has been reported [8,18] for a $\dot{V}_E$ in excess of 120 litres per min, i.e. ~15% of the current total V_{O_2}. This is presumably even greater in athletes, who are capable of attaining appreciably higher levels of $\dot{V}_E$ at maximum exercise.

Respiratory muscle perfusion ($\dot{Q}_{rm}$) naturally plays an important role in determining $\dot{V}_{rmO_2}$. Even though P_{aO_2} can decrease at high work rates in some highly trained athletes [19], arterial O_2 content (C_{aO_2}) is maintained essentially constant over the entire work-rate range. Consequently, the increased O_2 flux to the respiratory muscles depends on the increase in $\dot{Q}_{rm}$.

Neither the precise magnitude of the $\dot{Q}_{rm}$ increase during exercise, nor how it is partitioned among the respiratory muscles, has been rigorously established in humans. Using the $\dot{V}_{rmO_2}$ data of Shephard [16], and assuming that the arteriovenous O_2 content difference across the respiratory muscles increases to 15 ml per 100 ml at maximum exercise, Whipp and Pardy [15] estimated that $\dot{Q}_{rm}$ was 3.8 litres per min at maximum exercise, i.e. ~15% of a maximal cardiac output of 25 litres per min. This is remarkably similar to the values estimated by Dempsey and his associates [18,20] using more-direct methods. They determined that at a $\dot{Q}$ of 26.5 l/min the $\dot{Q}_{rm}$ was 4.2 l/min, i.e. ~16% of the total cardiac output.

This leads to the question of whether the O_2 demands of the respiratory muscles actually outstrip the vascular O_2 supply mechanisms during exercise. That is, do the respiratory muscles exercise beyond their 'lactate threshold'?

Specific respiratory-muscle endurance training [21–23] can increase not only $\dot{V}_{rmO_2}$ and respiratory muscle endurance, but also exercise tolerance [24], in normal subjects. It is not known, however, what proportion of this increased $\dot{V}_{rmO_2}$ is attributable to an increased $\dot{Q}_{rm}$, an improved O_2 extraction, or both.

Perfusion costs

Cardiac power appears to increase as a linear function of cardiac output during exercise in normal subjects [25]. The cardiac muscle power has been estimated to be ~30 times greater than the respiratory muscle power but at maximum exercise the respiratory muscle power may be greater than the cardiac power by a factor of three [25].

This suggests that the myocardial O_2 is likely to be a trivially small component of the total whole-body V_{O_2}, even at the high maximal cardiac outputs that can be achieved by highly fit athletes.

▶ Athletes capable of attaining high metabolic rates require commensurately high levels of ventilation during exercise.

▶ The high ventilatory demands of exercise in athletes require high levels of respiratory muscle work; this adds significantly to the total work associated with performing the task.

▶ This results in high metabolic and perfusion costs for the respiratory muscles, accounting for ~15% of the total oxygen uptake and cardiac output in fit, although not elite, athletes.

Constraints and limitations on respiratory system performance

The mechanical and metabolic costs of ventilation during muscular exercise are themselves sources of constraint and potential limitation. In the context of the respiratory system, constraint refers to a condition in which the ventilatory response is less than that required, owing to the influence of an opposing mechanism, for example an added resistive load. The system in this case is not actually limited from further increases in $\dot{V}_E$, as higher work rates lead to greater hyperpnoea.

In contrast, limitation occurs when the variable is actually prevented from increasing, despite further increases in ventilatory drive. Maximum expiratory air flow during exercise (at a particular lung volume) becomes limited despite further increases in ventilatory drive in subjects with reduced elastic recoil of the lungs and/or increased airways resistance; the achievable tidal volume is limited during exercise in patients with diffuse interstitial fibrosis as a result of the increased elastance (i.e. decreased lung compliance).

Ventilatory constraints

The tidal volume (V_T) limit in humans theoretically extends to the vital capacity and the breathing frequency up to ~5–7 Hz [26]. Normally, however, the ventilatory system 'only' operates at a frequency of ≤ 1 Hz and a V_T of only ~50–60% of the vital capacity, even during maximum exercise. The maximum airflow that can be attained at a given lung volume

during exercise in normal subjects is that generated by a maximal forced volitional effort, i.e. the maximum expiratory flow–volume (MEFV) relationship. During maximum exercise the spontaneously generated expiratory flow profiles fall well below the maxima of the MEFV curve [8,27] in subjects of poor or moderate fitness. However, in highly fit athletes, who are capable of generating high instantaneous airflows at the high levels of $\dot{V}_E$, these maxima may be encroached upon [8,15,28].

It was originally postulated by Rohrer [29] that respiratory work would be minimized at a particular level of $\dot{V}_E$, with a particular breathing frequency. A given $\dot{V}_E$ which is achieved with a high frequency and a low V_T would increase the flow-resistive component of the respiratory work. When this level of $\dot{V}_E$ is accomplished with a large V_T and a low breathing frequency, there would be an increased contribution from the elastic component of the respiratory work [30], as the lung volume encroaches on the flatter portion of the lung compliance curve. Exercising subjects seem to 'choose' a breathing pattern at, or near, the optimum for minimum respiratory muscle work rate [30] unless this is prevented by the breathing demands of the event, e.g. swimming.

There is likely to be turbulent constraint of airflow during exercise even in moderately fit subjects. It has been demonstrated that when the nitrogen fraction of the inspired air is replaced by the lower-density gas helium (care being taken to mask the sudden sensation of cold in the airways), there is no discernible effect on $\dot{V}_E$ at low levels of ventilation. However, at high levels of ventilation (compatible with significant turbulent airflow), replacing the nitrogen with helium induces a prompt and sustained hyperventilation. This is consistent with the removal of a constraint on $\dot{V}_E$ imposed by the turbulent flow. A similar result occurs if an added resistive load is suddenly removed, although in this case $\dot{V}_E$ increases because the hypoventilation, which was induced by the resistance, is removed.

However, the extent to which such turbulent constraint influences the respiratory compensation for the metabolic acidaemia of high-

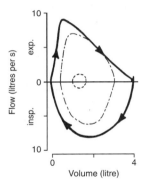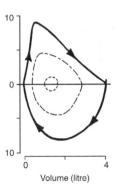

Fig. 6. Spontaneous F–V curves generated at rest (inner loop) and maximal exercise (middle loop) in a fit subject (left panel) and a normal, untrained subject (right panel), with reference to their maximum F–V curves (outer loop).

See text for discussion.

intensity exercise in athletes is not clear at present.

Ventilatory limitations

Mechanical

Ventilation appears to be neither mechanically limited nor associated with significant respiratory muscle fatigue during maximal exercise in moderately fit young individuals. For example: (i) $\dot{V}_E$ can be increased volitionally to levels appreciably greater than would be achieved spontaneously at maximum exercise; (ii) the ratio of maximum exercise $\dot{V}_E$ to MVV ($\dot{V}_E$/MVV) is relatively low (i.e. ~60–70%); (iii) the spontaneously generated expiratory flow–volume (F–V) curve does not encroach on the boundaries of the MEFV curve, even at maximum exercise; and (iv) there is also no evidence of significant respiratory muscle fatigue.

In subjects who are more fit, however, there is evidence of both ventilatory mechanical limitation and inspiratory muscle fatigue. It has been reported that the spontaneous expiratory F–V curve impacted on the outer envelope of the MEFV curve during maximal exercise in subjects with a $\mu\dot{V}_{O_2}$ of ~5–6 litres per min and a maximum exercise $\dot{V}_E$ of ~110–160 litres per min (i.e. some 80% of MVV; Fig. 6). It appears, however, that even with the evidence of airflow limitation such subjects generate only sufficient pleural pressures to establish the maximum flow [8,27,28], i.e they do not generate non-productive, airway-compressive pressures which would increase respiratory work without further benefit to airflow. During MVV manoeuvres, however, subjects do generate appreciably greater (and 'wasteful') pressures [8]. Furthermore, evidence of diaphragmatic fatigue has been demonstrated in such subjects, both in terms of electromyographic criteria and also as reduced maximum transdiaphragmatic pressures after exhausting exercise. Using the resting MEFV curve — and even the resting MVV — as the frame of reference for deciding whether there is airflow limitation during exercise should only be done with caution for the following reasons.

(i) Interpretation is crucially dependent upon appropriately 'locating' the spontaneous expiratory F–V curve on the MEFV curve. The better technique is to 'trap' an F–V display oscillographically on a particular breath and then perform the MEFV manoeuvre immediately thereafter, i.e. the MEFV manoeuvre is performed during the exercise.

(ii) It has been demonstrated that bronchodilatation can occur during exercise consequent to the increased circulating levels of catecholamines. This presumably accounts for the exercise MVV being greater that the resting MVV in some normal subjects.

(iii) The maximum expiratory effort F–V manoeuvre does not always yield the optimum maximum expiratory flow F–V curve — although in normal subjects this is more likely to be a factor in older athletes with diminished lung recoil and airways function than in healthy young athletes. In patients with chronic obstructive lung disease this is a particular problem.

There have been reports of reduction of vital capacity, respiratory muscle strength and endurance after exercise but this has only been consistently observed following prolonged exercise in athletes [31,32].

In summary, an athlete is confronted with a control dilemma at high work rates: to provide respiratory compensation for the metabolic acidosis (i.e. to constrain the pH fall in blood and exercising muscle), the additional $\dot{V}_E$ necessary can be so large (as described earlier) that it predisposes to ventilatory mechanical limitation and possibly ventilatory muscle fatigue. The absence of this compensatory hyperventilation tends to protect the subject from pulmonary mechanical limitation but at a cost of arterial hypoxaemia and a significantly greater fall in blood and muscle pH.

Metabolic

The respiratory muscles themselves have high metabolic demands during maximum exercise in highly fit athletes. $\dot{V}_E$ is not only high in such subjects at maximum exercise but the relationship between $\dot{V}_{rmO_2}$ and $\dot{V}_E$ becomes steep in this $\dot{V}_E$ range, as discussed above. Consequently, the blood flow that supplies this O_2 for utilization by the respiratory muscles is not available to supply the task-performing units. A greater component of the systemic O_2 supply is, therefore, 'diverted' away from the exercising limb musculature at these high levels of $\dot{V}_E$.

The theoretical limiting level of $\dot{V}_E$, above which the O_2 requirement of the respiratory

muscles becomes sufficiently large that it requires the entire further increment in whole-body $\dot{V}_{O_2}$, has been estimated to be in the range of 120–160 litres per min, i.e. levels of $\dot{V}_E$ which are not uncommon in elite athletes. Should the energetic requirements of the exercising limb musculature be preferentially met, the muscles of respiration would be predisposed towards fatigue. Conversely, if the requirements of the respiratory muscles are to be met entirely, the limb muscles would be predisposed to premature fatigue. During moderate exercise, both sets of muscles are likely to receive increased flow, as determined by the demand of the work rates being performed. During maximum exercise, however, Harms et al. [20] have reported a reduction in exercising-limb blood flow consequent to the demand for increased flow to the respiratory muscles.

Perfusion-related limitation

At all work rates above the lactate threshold (θ_L) $P_{A_{O_2}}$ increases progressively in normal subjects as a result of the additional ventilatory drive that normally increases $\dot{V}_E$ relative to $\dot{V}_{O_2}$. Why, therefore, is there not a systematic hyperoxaemia? Why does $P_{a_{O_2}}$ not increase in concert? In fact, the alveolar (ideal)–arterial P_{O_2} difference $([A–a]O_2)$ can increase to 20–30 mmHg, or more, at high work rates.

It has been widely demonstrated that moderately fit subjects at sea level maintain arterial P_{O_2} at, or close to, resting values even during exhausting work rates. For this to occur, the mean $\dot{V}_A/\dot{Q}$ for the entire lung must increase. This is because:

$$\overset{\text{Gas phase}}{\dot{V}_A \cdot (F_{i_{O_2}} - F_{A_{O_2}})} = \dot{V}_{O_2} = \overset{\text{Blood phase}}{\dot{Q}(C_{a_{O_2}} - C_{\bar{v}_{O_2}})}$$

$$(13)$$

and therefore

$$\frac{\dot{V}_A}{\dot{Q}} = \frac{C_{a_{O_2}} - C_{\bar{v}_{O_2}}}{F_{i_{O_2}} - F_{a_{O_2}}}$$

$$(14)$$

(We have disregarded, for simplicity, the small effect of the different inspired and expired tidal

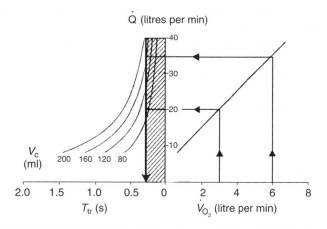

Fig. 7. Factors influencing the red blood cell transit through the pulmonary-capillary bed during exercise

Right panel: schematic representation of the relationship between cardiac output ($\dot{Q}$) and $\dot{V}_{O_2}$. Left panel: schematic representation of the relationship between $\dot{Q}$ and the mean pulmonary-capillary transit time (T_{tr}), with superimposed isopleths for pulmonary-capillary blood volume (V_c). The solid vertical line indicates a critical transit time of 0.3 s.

volumes that occurs when R $\neq$ 1, as it does not materially affect the argument.) $C_{\bar{v}O_2}$ decreases systematically whereas F_{AO_2} remains relatively constant as work rate increases in the moderate-intensity range: $\dot{V}_A/\dot{Q}$ therefore increases.

The distribution of $\dot{V}_A$ and $\dot{Q}$ has been shown to improve topographically during upright exercise. It was, therefore, assumed that the local distribution of alveolar ventilation-to-perfusion ($\dot{V}_A/\dot{Q}$) was also improved. The multiple inert gas technique of Wagner and associates [33], however, has failed to demonstrate an improvement in the dispersion of $\dot{V}_A/\dot{Q}$ in normal subjects during exercise. In fact, the dispersion of $\dot{V}_A/\dot{Q}$ within the lung has been shown to increase and this represents a significant component of the widening of the $[A-a]O_2$ that is normally seen during moderate and heavy exercise. This seems to be especially true at high work rates; neither Hammond et al. [34] nor Gale et al. [35] was able to detect statistically significant changes in $\dot{V}_A/\dot{Q}$ dispersion at more-moderate work rates ($\dot{V}_{O_2}$ <~2 litre per min), although Gledhill and co-workers did [36]. Hammond et al. [34] argued further that the increased $\dot{V}_A/\dot{Q}$ inequality did not entirely

account for the widened $[A-a]O_2$ at the high work rates (i.e. $\dot{V}_{O_2}$ × 3 litres per min).

At high work rates, diffusion impairment has also been implicated [34,35] as an additional contributor to the widened $[A-a]O_2$.

Dempsey et al. [19] demonstrated that P_{aO_2} actually fell at high work rates in some highly fit subjects. They suggested that diffusion impairment was contributory consequent to the high pulmonary blood flow. That is, the pulmonary capillary transit time (T_{tr}) was too short to allow diffusion equilibrium to be attained. Furthermore, Hammond et al. [34] found that neither the increased dispersion of $\dot{V}_A/\dot{Q}$ nor the increased post-pulmonary shunt was sufficiently large to account entirely for the widening of $[A-a]O_2$ at this high work rate in highly fit subjects exercising in 'the steady state' at 300 W.

The equilibrium of pulmonary end-capillary P_{O_2} with the alveolar P_{O_2} depends both on the rate at which P_{O_2} rises in the capillary bed and the time available for exchange. Approximately ~0.3 s is normally required for the P_{O_2} (and hence haemoglobin saturation) of the blood traversing the pulmonary-capillary bed to come into equilibrium with the alveolar gas partial pressure; this will be somewhat longer during exercise, however, especially in trained athletes, since the starting point (i.e. $P_{\bar{v}O_2}$) for the increase in capillary P_{O_2} is so low. This will be offset to some extent by the increased P_{AO_2}, which is normally seen at these high work rates and which increases the driving pressure for O_2 exchange.

If the flow through the capillary bed is sufficiently great, the blood will leave the gas-exchanging region and pass into the pulmonary veins before equilibrium occurs: the inevitable consequence is hypoxaemia, which becomes progressively greater the higher the flow. The residence time of blood in the pulmonary-capillary bed depends both on the capillary volume (V_c) and the pulmonary blood flow ($\dot{Q}$), i.e.

$$T_{tr}(s) \quad = \frac{V_c\,(\text{ml})}{\dot{Q}\,(\text{ml per s})} \tag{15}$$

where T_{tr} is the mean transit time.

The athlete must, therefore, recruit capillary volume as the capillary flow increases, to prevent the pulmonary-capillary residence time falling to critical levels. Each subject obviously has an upper limit to the capillary dimensions that is not discernibly increased by training — at least in the mature lung.

The relationship between T_{tr} and V_c is shown in Fig. 7. Note that the critical transit time of 0.3 s would be achieved at a $\dot{V}_{O_2}$ of 'only' 2 litres per min were there to be no increase in the resting $\dot{V}_c$ of ~80 ml. If the capillary volume doubles to 160 ml, the critical transit time would occur at a $\dot{V}_{O_2}$ of 'only' 5 litres per min. Further complicating this issue is the fact that the transit time is likely to vary (possibly widely) in different regions of the lung. As the pulmonary-capillaries do not all have the same lengths and diameters, there is, therefore, also likely to be a distribution of transit times. Transit times which are longer than the critical do not increase oxygenation further — however, the shorter ones do reduce it further. The actual flow-dependent hypoxaemia will then

begin to be manifest at work rates lower than suggested by the mean transit time.

A useful rule-of-thumb to determine the minimum pulmonary capillary volume (ml) that will maintain the mean transit time at 0.3 s is to multiply the cardiac output (litres per min) by 5:

$$\dot{V}_c(T_{tr} = 0.3) = 5 \times \dot{Q} \tag{16}$$

For a mean transit time of 0.6 the multiplier is 10.

The arterial hypoxaemia that has been demonstrated at high work rates in some fit athletes actually limits exercise performance. Exercise tolerance increased significantly [37] when the hypoxaemia was prevented by increasing the inhaled O_2 fraction sufficiently to maintain P_{aO_2}.

Why do all subjects exercising at high levels of pulmonary blood flow not develop arterial hypoxaemia? The answer is likely to be that some athletes have genetically large capillary beds — consequently, it may no longer be sufficient for an athlete to be 'elite' solely in the metabolic and cardiovascular senses. The appropriate pulmonary capabilities are also likely to be needed to permit the high metabolic rates necessary for future record performances in athletic events.

Genetics are likely to be dominant in this regard, since physical training does not systematically improve these indices of pulmonary function — although this notion is largely based on studies in adults, i.e. when the lung is mature. Whether these pulmonary dimensions can be improved by training during the period of lung growth in humans remains to be determined. The large pulmonary dimensions in some athletes who do train hard at times when the lung is not fully mature (e.g. swimmers) is suggestive, but not conclusive.

The high pulmonary vascular pressures associated with high pulmonary blood flow during exercise in elite athletes also predispose to both increased pulmonary-interstitial oedema and the potential for structural damage, including possible localized disruptions to the extremely thin alveolar–capillary interface [38]. This may prove to be of concern in some humans (as for the thoroughbred racehorse for which such disruption of the interface appears common). The driving pressure for blood flow across the lung is the difference between the pulmonary artery pressure and the pulmonary venous, or left atrial, pressure. The latter pressure is commonly estimated as the pulmonary 'wedge' pressure: a catheter is wedged into a small pulmonary artery and hence is only (at least ideally) influenced by the downstream pressure. Pulmonary artery pressures in excess of 40 mmHg have been reported at high levels of exercise (cardiac output ~25 litres per min) in healthy young subjects [39]. Pulmonary wedge pressures were seen to be 25–30 mmHg under these conditions. Similarly, high pulmonary vascular pressures were observed at appreciably lower levels of cardiac output in older [39], but healthy, subjects.

Whether there is net fluid flux (J) across the alveolar–capillary membrane will depend upon the difference between the hydrostatic pressure (tending to 'push' water out of the vascular bed) and the osmotic pressure (tending to 'pull' it back), i.e.

$$J = K \cdot (P_c - P_i) - K \cdot \sigma \cdot (\Pi_c - \Pi_i) \qquad (17)$$

where K is the filtration coefficient, P_c and Π_c are, respectively, the vascular hydrostatic and colloid osmotic pressures in the capillary, and P_i and Π_i are these pressures in the pulmonary interstitial (i.e. peri-microvascular) space. The term σ is the 'reflection coefficient', which is an index of the degree of the membrane impermeability to protein.

The behaviour of many of these variables during exercise in humans is poorly understood. Both the transpulmonary microvascular pressure and the area of the perfused surface increase during exercise; there is, however, no evidence of impaired capillary permeability. Factors that operate to prevent pulmonary oedema from developing during exercise are: (i) increased lung lymph flow; (ii) any fluid which

does move into the interstitial space will dilute its protein concentration and hence lower its osmotic pressure; and (iii) the interstitial space has a low compliance — consequently, small changes in volume will lead to large increases in peri-microvascular pressure. Both O'Brodovich and Coates [40], and Brower and Permutt [41], have concluded that these safety mechanisms are quite adequate to prevent the development of pulmonary oedema in normal humans — the evidence which does suggest the development of interstitial oedema during severe exercise is both indirect and inconclusive. Whether or not this occurs in highly trained athletes needs to be determined conclusively, because if it does occur during severe exercise it would predispose the athlete not only to further hypoxaemia, but also to pulmonary 'J' receptor stimulation. The latter would lead to greater tachypnoea, possibly increased respiratory sensation, and could even limit exercise tolerance further by inhibition of spinal motor neurons [42].

The issue of whether there may be stress failure of the fine-structured pulmonary capillaries during high-intensity exercise in these athletes at the high levels of capillary pressure has been explored by West and associates [38,43]. They demonstrated that short-term, high-intensity muscular exercise led to significantly increased concentrations of red blood cells, total protein and leukotriene B_4 in broncho-alveolar lavage fluid in elite cyclists but not in sedentary controls. Because they could demonstrate no difference between the exercising control group and the elite cyclists for tumour necrosis factor bioactivity, lipopolysaccharide or interleukin-8, they concluded that the changes in the athletes were likely to be a consequence of mechanical stress failure rather than being a primary inflammatory response. Interestingly, however, their more recent studies have been unable to demonstrate increases in red blood cells or leukotriene B_4 in the broncho-alveolar fluids of such elite cyclists who had exercised for 1 h at 'only' 80% of $\dot{V}_{O_2max}$. Therefore, stress-related failure of the pulmonary blood–gas interface in elite athletes, when it occurs, may only result from supramaximal efforts.

▶ Maximum attainable airflows can result during spontaneous breathing in some atheletes, i.e. their breathing becomes flow-limited.

▶ Whether this actually occurs in a particular athlete depends on the genetically determined structural make-up of the lung, i.e. its airway dimensions and elastic-recoil properties.

▶ The high levels of cardiac output (pulmonary blood flow) achieveable by athletes can propel red blood cells through the pulmonary-capillary bed before they have been fully oxygenated; arterial hypoxaemia (reduced P_{O_2}) results from this diffusion impairment.

▶ Whether this actually occurs in a particular athlete depends on the genetically determined structural make-up of the lung, i.e. its capillary properties.

▶ The gas-exchange efficiency of the lungs has been shown to be further compromised at high work rates by the regional matching of the alveolar ventilation to perfusion becoming less effective.

▶ The high pulmonary vascular pressures associated with the high pulmonary blood flow in athletes predisposes them to the development of interstitial oedema and even physical disruption of the fine-structured alveolar-capillary membrane. There is some evidence that this can actually occur, but when it does it is only at extremely high work rates.

Conclusions

▶ Healthy young subjects typically have pulmonary reserves even during exhausting exercise; consequently, exercise appears not to be limited by lung function.

▶ Athletes who are capable of attaining high levels of metabolic rate, with the associated high demands for both air and blood flow, are significantly more susceptible to pulmonary limitation. This is especially the case in elderly, but highly fit, subjects in whom deteriorations in pulmonary function have occurred as a seemingly inexorable consequence of the aging process.

▶ The ability to utilize the full metabolic and cardiovascular potential of an athlete appears to be dependent upon a permissive role of the subject's pulmonary structure, e.g. airway and pulmonary capillary dimensions.

▶ As this seems not to be altered by training (at least in the mature lung), the genetic make-up of the athlete's pulmonary system may prove to be of great importance in determining whether the exercise performance is impaired by limitation of the pulmonary system response directly or through its consequent influence on the functioning of other, related, physiological systems.

▶ Future record-breaking in sporting events that demand high levels of metabolic rate may, therefore, be restricted to those athletes with the appropriate genetically selected pulmonary systems.

Further reading

Johnson, B.D. and Dempsey, J.A. (1991) Demand vs. capacity in the aging pulmonary system. *Exercise Sports Sci. Rev.* **19**, 171–210

Whipp, B.J. (1998) Breathing during exercise. In *Pulmonary Diseases and Disorders* (Fishman, A.P., ed.), pp. 229–241, McGraw-Hill Inc., New York

References

1. Rahn, H. and Fenn, W.O. (1955) *A Graphical Analysis of the Respiratory Gas Exchange: The O_2–CO_2 Diagram*, American Physiological Society, Washington
2. Taylor, R and Jones, N.L. (1979) The reduction by training of CO_2 output during exercise. *Eur. J. Cardiol.* **9**, 53–62
3. Whipp, B.J. (1981) The control of the exercise hyperpnea. In *The Regulation of Breathing* (Hornbein, T., ed.), pp. 1069–1139, Marcel Dekker, New York
4. Whipp, B.J. and Ward, S.A. (1980) Ventilatory control dynamics during muscular exercise in man. *Int. J. Sports Med.* **1**, 146–159
5. Wasserman, K., Whipp, B.J., Koyal, S.N. and Beaver, W.L. (1973) Anaerobic threshold and respiratory gas exchange during exercise. *J. Appl. Physiol.* **35**, 236–243

6. McIlroy, M.R., Marshall, R. and Christie, R.V. (1954) The work of breathing in normal subjects. *Clin. Sci.* **13**, 127–136

7. Milic–Emili, G., Petit, J.M. and Deroanne, R. (1962) Mechanical work of breathing during exercise in trained and untrained subjects. *J. Appl. Physiol.* **17**, 43–46

8. Klas, J.V. and Dempsey, J.A. (1989) Voluntary versus reflex regulation of maximal exercise flow:volume loops. *Am. Rev. Respir. Dis.* **139**, 150–156

9. Leaver, D.G. and Pride, N.B. (1971) Flow–volume curves and expiratory pressures during exercise in patients with chronic airways obstruction. *Scand. J. Resp. Dis.* **77**, 23–27

10. Wasserman, K., Hansen, J.E., Sue, D.Y. and Whipp, B.J. (1987) *Principles of Exercise Testing and Interpretation*, Lea and Febiger, Philadelphia

11. Otis, A.B., Fenn, W.O. and Rahn, H. (1950) Mechanics of breathing in man. *J. Appl. Physiol.* **2**, 592–607

12. Jones, N.L., Killian, K.J. and Stubbing, D.G. (1988) The thorax in exercise. In *The Thorax* (Roussos, C. and Macklem, P.T., eds.), pp. 627–662, Marcel Dekker, New York

13. Grimby, G., Goldman, M. and Mead, J. (1976) Respiratory muscle action inferred from rib cage and abdominal V–P partitioning. *J. Appl. Physiol.* **41**, 739–751

14. Otis, A.B. (1964) The work of breathing. In *Handbook of Physiology: Respiration*, vol. 1 (Fenn, W.O. and Rahn, H., eds.), pp. 463–476, American Physiological Society, Washington

15. Whipp, B.J. and Pardy, R. (1986) Breathing during exercise. In *Handbook of Physiology: Respiration, Pulmonary Mechanics* (Macklem, P.T. and Mead, J., eds.), pp. 605–629, American Physiological Society, Washington

16. Shephard, R.J. (1966) The oxygen cost of breathing during vigorous exercise. *Q. J. Exp. Physiol.* **51**, 336–350

17. Karetsky, M.S. and Cain, S.M. (1972) Factors controlling O_2 uptake. *Chest* **61**(suppl.), 48S–49S

18. Aaron, F.A., Seow, A.C., Johnson, B.D. and Dempsey, J.A. (1992) Oxygen cost of exercise hyperpnoea: implications for performance. *J. Appl. Physiol.* **72**, 1818–1825

19. Dempsey, J.A., Hanson, P. and Henderson, K. (1984) Exercise-induced alveolar hypoxemia in healthy human subjects at sea level. *J. Physiol.* **355**, 161–175

20. Harms, C.A., Wetter, T.J., McClaran, S.R., Pegelow, D.F., Nickele, G.A., Hanson, P. and Dempsey, J.A. (1998) Effect of respiratory muscle work on cardiac output and its distribution during maximal exercise. *J. Appl. Physiol.* **85**, 609–618

21. Leith, D.E. and Bradley, M. (1976) Ventilatory muscle strength and endurance training. *J. Appl. Physiol.* **41**, 508–516

22. Belman, M.J. and Mittman, C. (1980) Ventilatory muscle training improves exercise capacity in chronic obstructive pulmonary disease patients. *Am. Rev. Respir. Dis.* **121**, 273–280

23. Folinsbee, L.J., Wallace, E.S., Bedi, J.F. and Horvath, S.M. (1983) Respiratory patterns and control during unrestrained human running. In *Modelling and Control of Breathing* (Whipp, B.J. and Wiberg, D.M., eds.), pp. 205–212, Elsevier, New York

24. Boutellier, U. (1998) Respiratory muscle fitness and exercise endurance in healthy humans. *Med. Sci. Sports Ex.* **30**, 1169–1173

25. Johnson, Jr., R.L. (1989) Heart–lung interactions in the transport of oxygen. In *Heart–Lung Interactions in Health and Disease* (Scharf, S.M. and Cassidy, S.S., eds.), pp. 5–41, Marcel Dekker, New York

26. Otis, A.B. and Guyatt, A.R. (1968) The maximal frequency of breathing in man at various tidal volumes. *Respir. Physiol.* **5**, 118–129

27. Olafsson, S. and Hyatt, R.E. (1969) Ventilatory mechanics and expiratory flow limitation during exercise in normal subjects. *J. Clin. Invest.* **48**, 564–573

28. Grimby, G., Saltin, B. and Wilhelmsen, L. (1971) Pulmonary flow–volume and pressure–volume relationship during submaximal and maximal exercise in young well-trained men. *Bull. Physio-Pathol. Respir.* **7**, 157–172

29. Rohrer, F. (1925) Physiologie der Atmebewegung. In *Handbuch der Normalen und Pathogischen Physiologie* (Bethe, A.T.J., von Bergmann, G., Embden, G. and Ellinger, A., eds.), pp. 70–127, Springer Verlag, Berlin

30. Milic-Emili, G., Petit, J.M. and Deroanne, R. (1960) The effects of respiratory rate on the mechanical work of breathing during muscular exercise. *Int. Z. Angew. Physiol. Einschl. Arbeitsphysiol.* **18**, 330–340

31. Bye, P.T.P., Farkas, G.A. and Roussos, C. (1983) Respiratory factors limiting exercise. *Annu. Rev. Physiol.* **45**, 439–451

32. Warren, G.L., Cureton, K.J. and Sparling, P.B. (1989) Does lung function limit performance in a 24-hour ultramarathon? *Resp. Physiol.* **78**, 253–264

33. Wagner, P.D., Saltzman, H.A. and West, J.B. (1974) Measurement of continuous distributions of ventilation-perfusion ratios: theory. *J. Appl. Physiol.* **36**, 88–99

34. Hammond, M.D., Gale, G.E., Kapitan, K.S., Ries, A. and Wagner, P.D. (1986) Pulmonary gas exchange in humans during exercise at sea level. *J. Appl. Physiol.* **60**, 1590–1598

35. Gale, G.E., Torre-Bueno, J., Moon, R., Saltzman, H.A. and Wagner, P.D. (1985) Ventilation–perfusion inequality in normal man during exercise at sea level and simulated altitude. *J. Appl. Physiol.* **58**, 978–988

36. Gledhill, N., Froese, A.B. and Dempsey, J.A. Ventialtion to perfusion distribution during exercise in health. In *Exercise and the Lung* (Dempsey, J.A. and Reed, C.E., eds.), pp. 325–343, University of Wisconsin Press, Madison

37. Powers, S.K., Lawler, J., Dempsey, J.A., Dodd, S. and Landry, G. (1989) Effects of incomplete pulmonary gas exchange on O_2max. *J. Appl. Physiol.* **66**, 2491–2495

38. West, J.B. and Mathieu-Costello, O. (1992) Strength of the pulmonary blood–gas barrier. *Resp. Physiol.* **88**, 141–148

39. Reeves, J.T., Dempsey, J.A. and Grover, J.F. (1989) Pulmonary circulation during exercise. In *Pulmonary Vascular Physiology and Pathophysiology* (Weir, E.K. and Reeves, J.T., eds.), pp. 107–133, Marcel Dekker, New York

40. O'Brodovich, H. and Coates, G. (1991) Lung water and solute movement during exercise. In *Exercise: Pulmonary Physiology and Pathophysiology* (Whipp, B.J. and Wasserman, K., eds.), pp. 253–270, Marcel Dekker, New York

41. Brower, R. and Permutt, S. (1991) Exercise and the pulmonary circulation. In *Exercise: Pulmonary Physiology and Pathophysiology* (Whipp, B.J. and Wasserman, K., eds.), pp. 201–220, Marcel Dekker, New York

42. Paintal, A.S. (1970) The mechanism of excitation of type-J receptors and the J-reflex. In *Breathing: Hering-Breuer Centenary Symposium* (Porter, R., ed.), pp. 59–66, Churchill, London

43. Hopkins, S.R., Schoene, R.B., Henderson, W.R., Spragg, R.G., Martin, T.R. and West, J.B. (1997) Intense exercise impairs the integrity of the pulmonary blood-gas barrier in elite athletes. *Am. J. Respir. Crit. Care Med.* **153**, 1090–1094

10

The effects of aging on exercise capacity

E. Joan Bassey
School of Biomedical Sciences, University of Nottingham Medical School, Clifton Boulevard, Nottingham NG7 2UH, U.K.

Introduction

Some individuals retain their muscular strength and stamina well into old age, others deteriorate and lose the physical capacity for independent and interesting lifestyles. The reasons for this variation are increasingly important in an aging society, and possible ways of maintaining physical capacities in later life are receiving attention. Although there is, in general, a decline in physical capabilities with increasing age during adult life [1], the rates of change are not uniform over the years between either individuals, particular physical capabilities or muscle groups. One characteristic of later life is a much larger variation in physical capabilities than is found among young people. Women are, in general, weaker than men at all ages because of a smaller absolute body size and a smaller proportion of muscle. They are, therefore, more likely to be incapacitated by loss of physical capability in old age than men. However, the patterns of change with age and the interaction of other factors, such as health and physical activity levels, do not differ, in general, between the genders.

The true, irreversible, time-dependent changes are overlaid with a rising incidence of chronic disease, accumulating effects of trauma and steadily diminishing activity levels — for which it is difficult to make allowance (Fig. 1). There are a number of chronic diseases, which develop slowly, manifesting themselves in middle or old age, which contribute to impaired exercise performance. Three of the more obvious ones that have direct effects are coronary heart disease, chronic obstructive lung disease and arthritis. About 50% of those aged over 65 years suffer from osteoarthritis, and an unknown but large proportion of middle-aged people have some degree of coronary occlusion caused by atheromatous plaques, even if they do not have overt symptoms of angina. In addition, there is a multitude of other chronic problems which may lead to reduced activity levels even if they do not initially impair exercise capability in a direct way. It must be recognized that healthy groups which have been screened for participation in laboratory studies will not represent population profiles, nor will the population profiles which include the effects of disease be an informative guide to changes that are purely time-dependent. This leads to philosophical debate about definitions of aging.

Muscle strength increases with training and decreases with lack of use, so studies of age-related changes are easily confounded by changes in the customary use of the muscles in question. Age-related changes are modest or absent in the studies in which activity levels do not change. Objective assessments in cross-sectional studies confirm the generally reported decline in customary activity levels with age and find it to be associated with a decline in muscle strength. If loss of strength is due — at least in part — to lack of activity, then the loss is likely to be in part reversible. However, where changes in activity levels and strength go hand in hand, it is rarely possible to say which of the two is causing the other. There is a chicken and egg situation. It is against this background that the evidence for irreversible, time-dependent changes and potentially reversible, activity-dependent changes will be presented in this chapter.

153

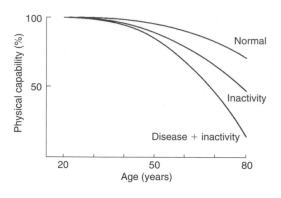

Fig. 1. Theoretical model of changes with age in physical capabilities with probable effects of disease and the additional effects of inactivity in both diseased and normal groups

Changes in muscle strength are of fundamental importance. If muscle strength declines, a loss of exercise performance of every kind will follow, even if the other components of the system remain unchanged. However, the other components are important as well. Variation in body fat, for example, affects weight-bearing activities such as walking. If the proportion of fat to muscle increases, the same muscle strength has to generate more force; so, for weight-bearing activities, there will be a loss of effective capability. Exercise performance may also depend crucially on energy supplies to the muscle, psychomotor skills and motivation, all of which may change for the worse with age, so age-related deficits may sometimes be additive in impairing functional performance. In addition, there are circumstances in which one particular deficit (such as motor control) may be critical; for example, stroke patients with high levels of spastic motor tone may be very strong but unable to walk.

This chapter has been split for convenience into subsections, but since such division is inevitably artificial there are also overlaps in places. It begins with the changes in muscle strength observed with age, considered in terms of the maximal force which a defined muscle group can apply briefly to some external measuring device, and then deals with the underlying changes in muscle cells responsible for producing that force. It progresses to muscle power, which is more functionally relevant than strength, since it is power or energy which people most frequently use as they move around. Power requires speed as well as force and can be assessed in brief maximal efforts, such as jumping and sprinting, or over prolonged periods of more moderate exercise, such as marathon running. The latter is usually assessed in terms of the oxygen consumed to do the work. Such prolonged exercise performance requires the support of the cardio-respiratory system which also suffers age-related changes; these will be presented briefly where they are relevant to exercise performance. Consideration of possible age-related changes in temperature regulation, energy balance and water balance may also be relevant but are beyond the scope of this chapter. Within each subsection, the effects of age on response to training have been considered, as have gender differences where appropriate.

Changes in muscle strength

Cross-sectional studies

Muscle strength increases rapidly during adolescence, stimulated by hormonal changes as well as skeletal growth, especially in males. During the years between full adulthood and incipient old age, muscle strength changes little, provided that physical activity levels are maintained. This is so for the various muscle groups which have been examined and for both isometric and isokinetic strength in men and women. Various studies of muscle strength in individuals between the ages of 25 and 65 years have found differences of between 0% and 50%. Within these studies there is a large variation, so that some of the 65-year-olds were stronger than some of the 25-year-olds, even after allowance was made for variation in body size. Results from a representative survey of 2700 men and women living in Britain aged between 25 and 65 years, who were assessed for isometric strength of the quadriceps and handgrip muscles [2], are shown in Fig. 2. The results

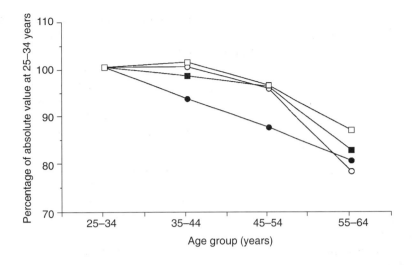

Fig. 2. Strength in quadriceps (circles) and handgrip muscles (squares) by age group

Data for 868 men (filled symbols) and 912 women (open symbols) adapted from a national survey [2]

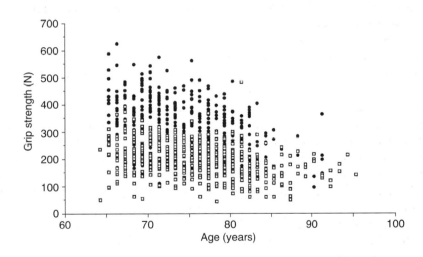

Fig. 3. Handgrip strength and age in 359 men (●) and 561 women (□) showing sex difference, decline with age and scatter in a representative population sample

Data taken from [4].

show that there is a mean loss of about 0.7% and 0.5% per year, respectively, in this population. Substantial proportions in all age groups had low activity levels and were weak in relation to the probable demands of a varied lifestyle; this was particularly so for women.

After 60 years, the decline is faster; for example, a rate of loss of muscle strength of 1.5% per year was described for 70-year-old men in Baltimore, U.S.A. [3]. Similar rates of loss after 65 years have been documented in several surveys in Britain in both men and women. In 1985, 1000 men and women aged 65–99 years, living in their own homes in a demographically representative area, were assessed for handgrip strength (see Fig. 3) and reported health and activity [4]. The mean handgrip strength for those aged 65 years was the same as those of the same age in the population survey conducted in 1990 and described in Fig. 2, confirming that the two studies were consistent. Body size, lack of health and lack of use of the handgrip muscles were significant associated factors. When these were taken into account, age still had a significant negative association, but it was smaller (about 1.0% per year). Women were much weaker than men at all ages.

Cross-sectional studies are vulnerable to cohort effects, so longitudinal studies are more valuable for assessing time-dependent changes.

Longitudinal studies

In a 5 year follow-up, in which 25 different muscle groups were measured in large samples of randomly selected men and women aged 70 years, mean decreases of between 1% and 3% per year were found [5]. However, the large standard deviations in the changes suggest that at least some of the subjects showed improvements over the 5 years, whereas others deteriorated. In other follow-up studies of isometric strength of handgrip and quadriceps muscles over 9 and 8 years, respectively, there was no change with age [3,5a]. Survivors in the group of 1000 elderly people described above were reassessed again after lapses of 4 and 8 years. The longitudinal changes in the first 4 years were found to be larger than predicted by the cross-sectional data, namely losses of 3% per year in men and nearly 5% per year in women [4]. Moreover, the loss was significantly related to age, confirming an accelerating loss of strength in old age. There was also a decline in health, physical activity and use of the muscles, but no change in body mass. The loss of use of the handgrip muscles was the dominant explaining variable for loss of strength; it was the only variable examined which followed the same pattern of accelerating loss as strength and it was not explained by deteriorating health.

Those who survived for 8 years showed the same pattern of association between loss of handgrip strength and reported use of the handgrip muscles [5b]. It cannot be assumed that the loss of use causes the loss of strength, rather than vice versa, but the evidence suggests that a significant part of the loss of strength with age may be attributable to lack of use of the muscles. There are also residual losses which constitute an inevitable time-dependent change.

Isokinetic strength

Many of the studies reported are of isokinetic strength. This should always be reported for a specific angle and measured late in the movement, especially at fast speeds, to allow adequate activation time and to avoid the inevitable impact artefacts of a mechanical over-ride system. Studies in which electrical stimulation is used to activate the muscle — instead of relying on the subject to produce maximal voluntary force — confirm that the loss of force with age is a real muscle phenomenon and not an artefact produced by the greater reluctance of an older person to recruit all available muscle fibres. Most of these studies are of force or torque recorded during concentric contractions. Recent studies of eccentric contractions, recorded while the muscle resists lengthening, show that the loss of force with age is proportionally less in eccentric than in concentric work [5c]. This was an unexpected finding which may be due to differences in collagen, which becomes stiffer with age.

Strength training

In old age, improvements in muscle strength, size and performance can be gained with training, following the usual rules which have been established in younger groups [6,7]. Improvements in both strength and area have been found with low-resistance, high-repetition training for 12 weeks in hamstrings and quadriceps in men aged over 70 years [8], in women [9,10] and in a very old group, aged 85–99 years [11]. Increases of 100% or more in 1RM (one repetition maximum) performance (a measure of dynamic strength in a defined movement) have been obtained. These are as large as those that can be obtained in younger individuals. The old can, therefore, be considered to be just as 'trainable' as the young. However, it must be remembered that there are also irreversible deficits in muscle bulk with age, so maximum physical potential is bound to be less [11a]. Some of the improvement is due to better neural control, but there is also a modest increase in the cross-sectional area of the muscle and of both slow- and fast-twitch fibres (see later) and the improvements are sometimes associated with relevant improvements in function, such as walking speed.

Amount of muscle

Large people are, in general, stronger than small people and the strength of a muscle depends on the cross-sectional area of its contractile elements. Body mass rises from youth to middle age, but this is due to the accumulation of body fat which is a burden in activities such as walking uphill. In old age there is a reduction in lean body mass which parallels the observed loss of strength, at least in broad terms, since lean body mass is mainly muscle. From 70 to 80 years of age, both muscle mass and cross-sectional area of the vastus lateralis muscle declined by 15% in men and women [12], which matches quite well the pattern they reported of loss of strength.

The reduced muscle mass is probably due to two processes. There is a gradual loss of muscle fibres caused by the death of motor neurons; this causes irreversible loss of the muscle fibre (apoptosis) unless it is re-innervated from a neighbouring motor neuron (see later). There is also a reduction in the size of those muscle fibres that remain, which is probably caused by lack of use and is therefore potentially reversible. The latter preferentially affects fast muscle fibres (which are larger) because these fibres are only recruited in rapid or vigorous movements which are not typical of the lifestyle of an old person. In a longitudinal study over 1 year of a group of steel workers after their retirement at 65 years of age in Britain, differential changes were found depending on the change in reported activity levels. A gain in muscle cross-sectional area was found in those

who reported increases in their activity levels and a loss in muscle cross-sectional area in those who reported a reduction in activity. This provides circumstantial evidence for a reversible component in the time-dependent loss of muscle bulk.

Measurements of gross muscle cross-sectional area must be interpreted with caution because they are not true measures of the cross-section of the contractile elements [13] but include connective tissue, water and fat. Within a muscle fibre there may be independent changes of contractile or other elements without a change in the diameter of the fibre. For example, after training the increase in muscle area is usually less than the increase in strength. Infiltration of the muscle with fat occurs with increasing age — and more so in women compared with men — so this can confound comparisons based on muscle cross-sectional area and renders futile most of the debate about age-related changes in the 'intrinsic' strength of a muscle, defined as the ratio of strength to cross-sectional area.

Muscle fibres

Explanations for the observed changes with age have been sought using needle biopsies from conscious volunteers. The samples are rapidly frozen and later sectioned and stained. This has allowed investigations of the proportions and distributions of the various fibre types in muscle, the cross-sectional area of the fibres, the concentrations of enzymes and mitochondria and the capillary density. The technique is powerful but the data need to be interpreted with caution because of the uncertainties of sampling a whole muscle with a single small muscle biopsy.

The cross-sectional area of both slow and fast fibres is well maintained until 70 years of age. After that the fast fibres become smaller [12] but some studies find that the slow fibres are still well maintained. This may be a result of a reduction with age in strenuous activities likely to recruit fast fibres and the maintenance in the groups studied of a range of modest activities which preserve slow fibre size.

Indirect estimates of the number of fibres in a muscle and age-related changes have been made by combining mean cross-sectional areas of the fibres found in biopsies along with total muscle cross-sectional areas from scans. The reductions found with age in the fibre areas cannot account for the decrease in the total cross-section of the whole muscle, at least not until after the age of 70 years. In addition, painstaking direct estimates have been made of the total number of fibres in cadaver muscles. Both approaches have led to the conclusion that a reduction in the total number of fibres occurs [14,15]. This is attributed to the irreversible death of motor neurons that occurs alongside the death of other nerve cells with increasing age. There is no good evidence for a selective loss of numbers of fast muscle fibres rather than slow fibres with increasing age.

These losses are offset to some extent by re-innervation of fibres which have lost their motor neuron. The evidence for this includes a change with age in the distribution of the muscle fibre types throughout a muscle, with increasing clumping of fibre types into groups of a similar type instead of the even dispersion seen in a young muscle; and also the finding that motor units are fewer but larger in old muscle than in young [16,17]. This may be associated with less precise control in graded contractions. There is a limit to how many fibres a given motor neuron can serve, and some fibres will therefore die. There is no evidence as yet that activity can influence the rate of loss of motor neurons or the rate or extent of re-innervation; these are possibilities that remain to be explored. It is probable that muscle fibres are very robust and intrinsically suffer little with the passage of time. The loss of muscle in very old age may be entirely a consequence of progressive degeneration of the central nervous system.

Inactivity reduces fibre size in subjects of all ages [18]. The extreme case of immobilization of a limb due to fracture can lead rapidly to a 50% reduction in mean fibre size. This is apparent, for example, when a footballer is confined to bedrest following a knee injury: the rate of shrinkage of the quadriceps bulk is noticeable without making formal assessment. Space research teams are still trying to find ways of prescribing exercises which would combat the disabling amount of muscle loss that occurs when their personnel are exposed for long periods to lack of the gravitational force against which muscles normally have to work. Strength training, on the other hand, can increase the size of all fibre types and this can happen even in old age [8]. With high-resistance, weight-lifting training for 12 weeks, increases of 30% in mean fast-fibre area have been achieved in men aged between 60 and 70 years [6]. In veteran cyclists, very large slow fibres with an abundance of mitochondria have been found in quadriceps. New muscle fibres (hyperplasia) probably do not occur, but it is agreed that hypertrophy (enlargement) of fibres can take place with a suitable stimulus even at advanced years.

The stimulus for hypertrophy has not yet been established but it is possible that stretching of the sarcolemma triggers activity of the nucleic acids in the cell (DNA and RNA) which leads to formation of new protein. This process is also triggered by minor damage to the sarcolemma, caused by the kind of strenuous physical activity associated with temporarily stiff and sore muscles. All muscle cells contain the genetic information needed for expressing a variety of types of contractile protein (different myosins) and so can potentially become fast or slow or intermediate fibres depending upon the neural traffic reaching them (a high rate of neural stimulation results in a fast muscle fibre, and a slow rate results in a slow muscle fibre). In old age, this plasticity is retained but there is also a change to slower forms of myosin.

The basement membranes between cells and around capillaries are seen to be thickened in old muscle and it has been suggested that this decreases the rate of diffusion for oxygen, energy substrates and metabolites, although there is no direct evidence that this happens to a significant extent. Benign inclusions of various kinds are found in old muscle cells, such as lipofuscin granules. They are thought to be lysosomal residual bodies. Their origin and significance are unknown, and they are perhaps

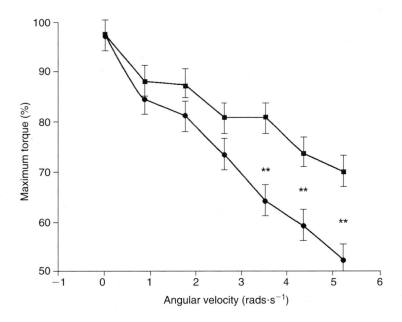

Fig. 4. Torque velocity relations for quadriceps in concentric contraction for an old (•) compared with a young (■) group of women with similar activity levels
**Indicates a significant difference between age groups (P<0.01). Data taken from [20].

best thought of as accumulated junk which causes nothing worse than clutter.

Speed of muscular contraction
Muscles become slower with age. The evidence for this is that when a muscle is stimulated directly, using a single maximal electrical shock, or indirectly, using a reflex pathway, the force produced rises to its peak more slowly in older muscle and it also relaxes more slowly. One consequence of the slower rise of tension is that a fused tetanus will be obtained at a lower fusion frequency than in young muscle. The old muscle behaves in this respect more like a muscle which is fatigued. It also means that a reduction in neural conduction velocity which has been reported in the very old would not, of itself, lead to reduced muscle function. The maximal voluntary speed of movement of the knee extensors falls with age [19], maximal short-term power output falls with age [2] and

studies of the force–velocity curves suggest that the decrement in force with speed is proportionally greater with increasing age in both men and women [20] (see also Fig. 4). Older individuals are, therefore, not only weaker in terms of isometric strength but progressively disadvantaged when they need to generate force rapidly.

Muscle endurance and fatigue
There are many ways to define endurance but it is usually measured as the maximum time for which some specified muscular performance can be continued. A commonly used definition relative to strength is the length of time for which an individual can maintain an isometric contraction at a known percentage of the previously measured maximal voluntary contraction (MVC) force. This amounts to 45–60 s at 50% MVC. Such maximal endurance performances, when taken to the limit, cause substantial rises in blood pressure which constitute a risk in

older individuals because of possible hidden weaknesses in the cardiovascular system. Absolute as well as relative levels of force are important here, since endurance time is thought to be a function of developing ischaemia due to the occlusion of muscle blood flow by the contraction. This results in weaker people having better endurance provided they are assessed relative to their own MVC force rather than at some absolute force level. No change has been found with age in endurance so defined. Indeed, older individuals and women may have unexpectedly good endurance just because they are weaker. This is, however, a somewhat artificial concept which is of little use when faced with a heavy suitcase and a long walk.

No change was found with age in a fatigue index in repeated voluntary dynamic effort with the quadriceps muscle [21]. This was confirmed for the triceps surae muscle using tetanic stimulation. Here again, the absolute efforts will be less in an old person, but relative to their own maximal strength they do not fatigue more easily than a young person. They may even fatigue less if their muscle fibres are composed of predominantly slow, fatigue-resistant fibres.

Hormones

There are changes throughout life in circulating hormone levels which may influence physical capabilities. Growth hormone levels may be low in elderly individuals and if so treatment may increase strength. Testosterone levels also fall with age and it has been suggested recently that the drop in oestrogen levels at the menopause may be associated with a drop in strength in women. However, randomized, double-blind trials of hormone replacement therapy show convincingly that there is no increase in strength with this treatment [22].

▶ Between 30 and 60 years of age, there is an average loss of muscle strength of about 20%, this is a mean rate of loss of less than 1% per year. The rate of loss then accelerates.
▶ There is much variation, because of the effects of lack of use and disease as well as inevitable time-dependent change.
▶ In old age, some may have strengths which are only just adequate for an independent lifestyle and which provide little safety margin for sudden demand; this is especially true for women.

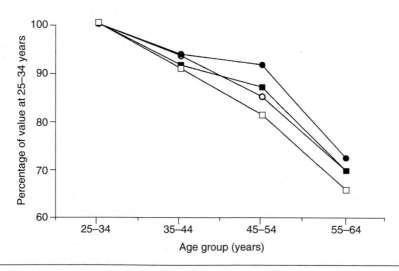

Fig. 5. Power output from the leg in rapid extension by age group derived from watts (round symbols) or from the ratio of watts per kg of body mass (square symbols) for men (closed symbols) and women (open symbols) adapted from a national survey [2]

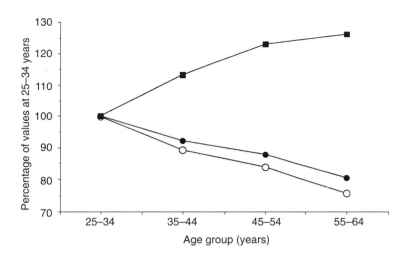

Fig. 6. Predicted maximal oxygen uptake from a sub-maximal treadmill test derived from data in litre/min (●) and in ml per kg of body weight per min (○); and percentage fat from skin-fold measurements (■); data for 682 women adapted from a national survey [2]

▶ The functional consequences of loss of strength, for weight-bearing activities such as walking, are greater if body fat stores have also increased.

▶ Muscle strength, muscle size and functional performance can all increase with training even in very old age.

Changes in power output

Short-term, rapid power output

Short-term, rapid power output is taken to include explosive power measured during jumping, or sprinting for up to 30 s on stairs, ramps or a cycle. Performance in these activities depends on both speed and force and the aging muscle suffers some inevitable reductions in both. Selective reduction in cross-sectional area of type-II (fast) fibres may explain some of the change. In addition, the contribution of stored elastic energy in the stretch-shortening cycles of rhythmic movements will be reduced as the resilience of collagen in tendons, ligaments and joints diminishes with age [23]. This may also explain why women walk less efficiently when they are older.

Most old people are not able to jump or sprint, so until recently these studies have been on unusually capable groups. There are master athletes in their 70s in many sports, and aged marathon runners, but no aged sprinters. Equipment for measuring leg extensor power in a single push has been developed for the safe testing of power in all age groups, including the frail elderly. The extensor push is made seated, without involving the back, and lack of balance is irrelevant. The extensor push starts and finishes at rest, and has properties in common with stepping up a stair or arising from a chair [24]. Using this set up, a decline in explosive power with age of about 11% per decade has been found in a large survey (see Fig. 5). No longitudinal studies are available but the cross-sectional data suggest that the rate of loss of short-term power with age is much greater than the rate of loss of strength but just as variable. Improvements in leg extensor power can be achieved with training, even in old age [24a,24b].

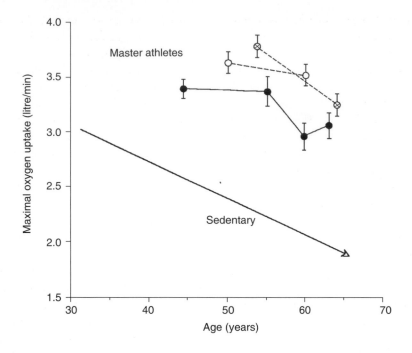

Fig. 7. Maximal oxygen uptake and age; data from two groups of men who continued running training, compared with average data from the literature for sedentary groups

Open symbols, data taken from [25]; closed symbols, data taken from [26]. Open symbols enclosing a × represent changes in a sub-group who ceased training.

Medium-term power output (maximal oxygen uptake)

The ability to produce power continuously for many minutes in rhythmic movements such as running or swimming depends on the throughput of aerobic energy of the muscle. Mitochondria and aerobic enzymes are needed within the muscle and an adequate flow of oxygen to them from the lungs via the central circulation and the local capillary bed. Performance is closely related to the maximal oxygen uptake capability of the whole system. If this is maximally challenged over a period of 6–8 min using the large muscle groups of the legs, as in running, then the limiting factor is thought to be the maximal cardiac output. Under these circumstances the total oxidative capacity of the active muscle bulk exceeds the capacity of the cardiovascular system to deliver blood flow while maintaining arterial pressure within the limits

dictated by homeostatic mechanisms. Exceptions to this might occur where muscles have become severely deconditioned. If small muscle groups are used, the limiting factor is the capacity of the muscle to receive blood flow and extract oxygen from it.

As external power output continues to increase, a plateau in the oxygen uptake is evidence that the maximal capacity of the oxygen-transport system has been taxed. At this stage, the energy supply is only partly aerobic, with an additional short-term anaerobic contribution of about 20%. The plateau criterion is rarely achieved in old subjects because symptoms of fatigue or signs of myocardial ischaemia occur first. Other, less clear-cut indices — such as high levels of lactate in arterialized blood or heart rate — are used as evidence that there is a decline in maximum oxygen uptake with age. The outcome then is a symptom-limited maximum which is functionally valid but may be telling us different things about limiting factors.

Findings from cross-sectional studies show that maximal oxygen uptake [2], defined in either of these ways, declines steadily after 25 years at about 0.7% per year (Fig. 6). The decline evidently begins earlier than the age-related decline in muscle strength and, therefore, cannot be entirely due to that loss. As with strength, the effects of declining activity will confound the effects of increasing age; and, again, in the studies which attempt to control for this, the loss with age is reduced sometimes to negligible levels. For instance, in a comparison between young (26 years) and middle-aged (56 years) runners, who were matched for training and performance, the older men had a maximal oxygen uptake (plateau criterion) which was only 9% less than the younger men. This represents a mean cross-sectional loss of only 0.3% per year, less than half of that observed in the population.

Longitudinal studies confirm the cross-sectional studies in general but there is much variation and it is clear that changing activity levels affect the rate of loss. Two studies of men who maintained an athletic lifestyle from middle age onwards (Fig. 7) showed that there was much less deterioration than in the general population over 10 years [25] and 25 years [26]. It is clear that healthy individuals who maintain activity levels suffer little deterioration in their maximal aerobic capacity at least up to 70 years or so.

Maximal oxygen uptake is sometimes expressed in ml per kg of body mass, rather than as absolute litre per min, to allow comparisons to be made across individuals of varying body size. This must be treated cautiously because it is only legitimate if the body composition, i.e. the ratio of fat to lean body mass, is similar. For example, a decline with age in maximal oxygen uptake expressed in ml per kg of body mass may indicate an increase in body fat rather than a reduction in aerobic capacity, since body fat tends to increase with age (Fig. 6). However, in either case, there is a decrease

in the power-to-weight ratio of the body for weight-bearing activities, such as walking, running or climbing stairs.

Because of the difficulties of measuring maximal oxygen uptake directly in old individuals, predictions from the linear relation between heart rate and oxygen uptake are sometimes made. This involves making assumptions about the individual's predicted maximal heart rate and can introduce considerable error. Since there is an inverse relationship between changes in maximal oxygen uptake and changes in sub-maximal heart rate at a standard level of oxygen uptake, the latter is taken as a useful index of these changes. This is a useful approach for assessing training effects within a subject but many studies show that there is no age-related change in sub-maximal heart rate despite reductions in maximal oxygen uptake, so it is less useful for comparing age groups. Two opposing changes occur: a reduction in maximal heart rate due to age and an increase in sub-maximal heart rates due to inactivity.

Convincing evidence has now accumulated from a number of controlled intervention studies in middle-aged and older subjects [27], showing not only that potential for improvement with training remains even in those over 70 years of age [28] but also that it can happen in 8 weeks [29]. Old groups appear to be just as trainable as young groups, if not more so, since they start from a more sedentary baseline. Walking is a sufficiently energetic activity to stimulate improvements in some old people and low-intensity training at 50% of predicted maximal intensity appears to be adequate for significant gains. As in young people, training can produce increases both in the oxidative capacity of the muscles used and in the maximal capacity of the central delivery system. The central improvements are due to changes in control mechanisms as well as increases in myocardial contractility or blood volume. Peripheral improvements are due to increases in the oxidative capacity of the muscle, and a more favourable (to the muscle) distribution of the cardiac output or local flow.

Prolonged power output (endurance)

Few normal individuals ever work at their maximal capacity and as age advances this becomes increasingly so; capacity for moderate prolonged work is well maintained with age. It is of interest to consider prolonged power output which can be sustained for more than an hour. This can also be defined as endurance, and identified as the maximum exercise intensity at which the muscles are doing work aerobically — sustained by adequate supplies of oxygen — and at which lactate levels do not rise much above resting levels; typically a long brisk walk. This exercise is at the lower boundary of a functional threshold; a small increase in exercise intensity will cause the muscle to depend on anaerobic as well as aerobic energy sources so that levels of lactate will begin to rise and symptoms of fatigue develop. This functionally significant departure point has been variously defined as the lactate threshold, the onset of blood lactate (OBLA), the aerobic threshold and the anaerobic threshold. The definition depends on the methodology: respiratory indices as well as levels of lactate in arterialized blood of 2.0 mM or 4.0 mM have been used (resting is less than 1.0 mM). All techniques have the problem of identifying a point on what some have shown to be a continuously rising curve. In addition, there are lactate fluxes both into blood from working muscles and out of the blood into resting muscles, heart and liver tissue which can use lactate as a fuel if there is enough oxygen around to metabolize it all the way to carbon dioxide. This leads to uncertainty about the lactate concentrations within the working muscle cells unless they are measured directly from biopsies.

The observation that many old people retain their ability for prolonged moderate performance is borne out by studies which have examined various versions of this functional threshold over a wide age range. The power output which can be reached and maintained for some time without raising lactate levels is sometimes well preserved with age; even if it is not, the loss is rather less than the reduction in the maximal transporting capacity for oxygen. When a middle-aged and young group of runners were matched for training and performance, the decrement of 9% in maximal oxygen uptake (plateau criterion) found in the older men was offset by a higher relative lactate threshold [30,31]. At 2.5 mM lactate, the two groups had the same running speeds and the same oxygen uptake.

The cardiovascular system

No change with age has been found in biopsy studies in the ratio of capillaries to muscle fibres, nor in the relative number and volume of mitochondria. The metabolic enzyme activity of the remaining muscle appears to be undiminished [12]. However, it should be borne in mind that sedentary lifestyles which are typical of old age are associated with poorer capillary-to-muscle fibre ratios. The converse is also true — at any age the ratio can improve with exercise.

A marked, linear but variable, decrease with age in maximal heart rate during rhythmic exercise is always found. It has not yet been satisfactorily explained; it has been attributed to loss of sympathetic motor neurons but there are alternative possibilities. There are studies which suggest that there are changes with increasing age in sympathetic outflow at rest, changes in autonomic receptor populations, changes in their sensitivity and changes in baroreceptor control; however, none of these supposed age-related changes has been substantiated. Some studies also show that an increase in maximal heart rate can occur with training in both young and old groups [28]; again, the mechanism is unknown.

Algorhythms have been published for calculating predicted maximal heart rate for a specified age. These are required because of the inconvenience or danger of direct measurement in some old subjects; however, the fact that there are several different algorhythms raises suspicions about their validity. There is considerable variation in measured maximal heart rates at any age. For instance, a maximal heart rate of 171 beats per min has been directly recorded in a man aged 71 years, although one of the predicted values for a man of this age is 124 beats per min. In young people, maximal rates of 215 beats per min are not uncommon but others produce a maximal plateau, despite increasing muscular work output, at 175 beats per min.

The reduction in maximal heart rate usually accompanies a decrease in maximal cardiac output. However, the true age-related decrease is now thought to be modest. Many early studies included subjects with undetected coronary artery occlusion. It is now possible to screen for this and a cross-sectional study of healthy volunteers has shown no change in maximal cardiac output up to the age of 79 years [32]. This was achieved with a larger maximal stroke volume and lower rate than in the younger subjects in this study. The larger stroke is accompanied by a larger end-diastolic volume, in contrast with the larger end-systolic volume of the well-trained athlete. The implications of this for myocardial work and safety margins are not clear. It is of interest that in this study there was no decrement in maximal, medium-term power output with age.

Maximal cardiac output can also increase by as much as 30% with training in old individuals [28]. This study was of a 12-week, high-intensity, endurance-exercise training programme, in two groups of men aged 60–70 years and 20–30 years, matched for height, weight and physical activity. Maximal oxygen uptake increased by 38% in the old group and by 29% in the young group, confirming the 'trainability' of older individuals. The increase in the old group was mainly due to a 21% increase in stroke volume. However, in contrast with the cross-sectional study described above, the older group had a cardiac output that was 25% less than the young group, even after training.

Blood pressure tends to rise with age in developed countries [33], but this is not a universal finding. It may be secondary to increased body fat or some other consequences of a too-affluent diet. It is of interest that in a longitudinal study of 12 runners no changes were found in resting heart rate, weight or resting blood pressure over 25 years, by which time the mean ($\pm$S.D.) age of the subjects was 69 ± 7 years [26]. Capillary branching remains abundant so that the diffusing distance from capillary to muscle

fibre remains short in a healthy old person who continues to exercise. The reverse occurs if that muscle is not used, just as in a younger person. Maximal muscle blood flow will fall with age if maximal cardiac output falls. Arteriosclerosis may impair the distribution of the cardiac output. If there is severe narrowing of the blood vessels supplying the muscle, such as happens in the leg muscles in a disease called intermittent claudication, then exercise capacity is severely impaired.

Respiratory system

There are marked irreversible changes in the lungs and respiratory system with age but, because the safety margins are so large in young healthy individuals, the losses do not appear to limit normal physical activities in healthy old people; ventilation rates of over 100 litre per min have been recorded during maximal exercise in subjects aged up to 72 years. An increase in ventilation during maximal work observed in training programmes may indicate adequate safety margins rather than respiratory improvements. However, for master athletes who, despite their age, are still performing at high level and have optimally trained muscles, these inevitable losses become important and may limit their oxygen uptake [34].

There is a decrease in vital capacity and peak expiratory flow rate even in healthy old lungs due mainly to loss of elastic recoil caused by emphysematous changes in the lung tissue. These changes also result in a smaller diffusing surface area. They are exacerbated by smoking or other forms of chronic atmospheric pollution. If there is disease — such as bronchitis, severe emphysema, fibrosis or lung cancer — then severe impairment may occur to the point where the patient feels breathless at rest and no exercise, even of the mildest kind, is possible.

The central nervous system, co-ordination and balance

Old people have a central nervous system containing fewer neurons; this leads to less-effective networking in the brain with longer processing times for complex tasks. There is an increase in the processing time required by the central nervous system and a decrease in sensitivity of many sensory nerve endings; this will affect physical capabilities to some extent and will reduce the capacity to acquire new physical skills. Balance and postural righting reactions become poorer with age; they require the rapid collation of visual and proprioceptive input, as well as input from the auditory canals, followed by a rapid motor response of adequate speed and power. There may be failure in any of the links of the chain, or an additive slowing of the links to a point where the whole reaction takes too long and a fall occurs.

► The reasons for the inevitable loss of physical capabilities with age and the selective loss of fast, powerful performance include loss of muscle mass, an intrinsic slowing of remaining muscle and a reduction in the maximal capacity of the oxygen-delivery system.

► There are changes in connective tissue, which becomes less resilient, and its capacity for storing and releasing energy in the stretch–shortening cycles of rapid movement is therefore diminished.

► Age-related changes in the different components of the system are not likely to be uniform, and there will be genetic variability. A number of small changes may have multiplicative rather than additive effects when maximal performance is considered and some deficits in the central nervous system may be crucial.

► Sub-maximal capability is often well maintained but the safety margins will be smaller.

► Activity patterns are important and, even in old age, substantial and functionally relevant improvements with training are possible. It is clear that maintained use of muscles in a variety of kinds of exercise will offset a good deal of the deterioration observed on average with increasing age in adults.

I would like to apologise to all those colleagues whose work I have not quoted due to lack of space.

Further reading

A report on activity patterns and fitness levels. (1992) In *Activity and Health Research: Allied Dunbar National Fitness Survey.* The Sports Council and Health Education Authority, London

Bassey, E.J. and Harries, U.J. (1993) Normal values for handgrip strength in 920 men and women aged over 65 years, and longitudinal changes over 4 years in 620 survivors. *Clin. Sci.* **84**, 331–337

Fiatarone, M.A., Marks, E.C., Ryan, N.D., Meredith, C.N., Lipsitz, L.A. and Evans, W.J. (1990) High intensity strength training in nonagenarians: effects on skeletal muscle. *J. Am. Med. Assoc.* **263**, 3029–3034

Grimby, G. and Saltin, B. (1983) The aging muscle: mini-review. *Clin. Physiol.* **3**, 209–218

Makrides, L., Heigenhauser, G.J.F. and Jones, N. (1990) High intensity endurance training in 20- to 30- and 60- to 70-yr-old healthy men. *J. Appl. Physiol.* **69**, 1792–1798

References

1. Grimby, G. and Saltin, B. (1983) The aging muscle: mini-review. *Clin. Physiol.* **3**, 209–218
2. A report on activity patterns and fitness levels. (1992) In *Activity and Health Research: Allied Dunbar National Fitness Survey* The Sports Council and Health Education Authority, London
3. Kallman. D.A., Plato, C.C. and Tobin, J.D. (1990) The role of muscle loss in the age related decline of grip strength: cross-sectional and longitudinal perspectives. *J. Geront.* **45(3)**, M82–M88
4. Bassey, E.J. and Harries, U.J. (1993) Normal values for handgrip strength in 920 men and women aged over 65 years, and longitudinal changes over 4 years in 620 survivors. *Clin. Sci.* **84**, 331–337
5. Aniansson, A., Sperling, L., Rundgren, K. and Lehnberg, E. (1983) Muscle function in 75-year-old men and women: A longitudinal study. *Scand. J. Rehabil. Med.* **9**(Suppl.), 92–102
5a. Greig, C., Botella, J. and Young, A. (1993) The quadriceps strength of healthy elderly people remeasured after 8 years. *Muscle Nerve* **16**, 6–10
5b. Bassey, E.J. (1998) Changes in selected physical capabilities; muscle strength, flexibility and body size. *Age Ageing* **27(3)**, 12–16
5c. Porter, M.M., Vandervoort, A.A. and Kramer, J.F. (1997) Eccentric peak torque of the planar and dorsiflexors is maintained in older women. *J. Geront.* **52A**, B125–B131
6. Brown, A.B., McCartney. N. and Sale, D. (1990) Positive adaptations to weight-lifting training in the elderly. *J. Appl. Physiol.* **69**, 1725–1733
7. Pyka, G., Lindenberger, E., Charette, S. and Marcus, R. (1994) Muscle strength and fiber adaptations to a year-long resistance training program in elderly men and women. *J. Geront.* **49**, M22–M27
8. Frontera, W.R., Meredith, C.N., O'Reilly, K.P., Knuttgen, H.G. and Evans, W.J. (1988) Strength conditioning in older men: skeletal muscle hypertrophy and improved function. *J. Appl. Physiol.* **64**, 1038–1044
9. Charette, S.L., McEvoy, L., Pyka, G., Snowharter, C., Guido, D., Wiswell, R.A. and Marcus, R. (1991) Muscle hypertrophy response to resistance training in older women. *J. Appl. Physiol.* **70(5)**, 1912–1916
10. Cress, M.E., Thomas, D.P., Johnson, J., Kasch, F.W., Cassens, R.G., Smith, E.L. and Agre, J.C. (1991) Effect of training on $\dot{V}_{O_2 max}$, thigh strength, and muscle morphology in septuagenarian women. *Med. Sci. Sports Exercise* **23(6)**, 752–758

11. Fiatarone, M.A., Marks, E.C., Ryan, N.D., Meredith, C.N., Lipsitz, L.A. and Evans, W.J. (1990) High intensity strength training in nonagenarians; effects on skeletal muscle. *J. Am. Med. Assoc.* **263**, 3029–3034

11a. Welle, S., Totterman, S. and Thornton, C. (1996) Effect of age on muscle hypertrophy induced by resistance training. *J. Geront.* **51A**, M270–M275

12. Grimby, G., Danneskiold-Samsoe, B., Hvid, K. and Saltin, B. (1982) Morphology and enzymatic capacity in arm and leg muscles in 78–81 year old men and women. *Acta Physiol. Scand.* **115**, 125–134

13. Forsberg, A.M., Nilsson, E., Werneman, J., Bergstrom, J. and Hultman, E. (1991) Muscle composition in relation to age and sex. *Clin. Sci.* **81**, 249–256

14. Lexell, J. and Downham, D.Y. (1992) What determines the muscle cross-sectional area? *J. Neurol. Sci.* **111**, 113–114

15. Sjostrom, M., Lexell, J. and Downham, D.Y. (1992) Differences in fibre number and fibre type proportion within fascicles. A quantitative morphological study of whole vastus lateralis muscle from childhood to old age. *Anat. Rec.* **234**, 183–189

16. Campbell, M.J., McComas, A.J. and Petito, F. (1973) Physiological changes in aging muscles. *J. Neurol. Neurosurg. Psych.* **36**, 174–182

17. Doherty, T.J., Vandervoort, A.A., Taylor, A.W. and Brown, W.F. (1993) Effects of motor unit losses on strength in older men and women. *J. Appl. Physiol.* **74**, 868–874

18. Kristensen, J.H., Hansen, T.I. and Saltin, B. (1980) Cross-sectional and fiber size changes in the quadriceps muscle of man with immobilization and physical training. *Muscle Nerve* **3(38)**, 275–276

19. Larsson, L., Grimby, G. and Karlsson, J. (1979) Muscle strength and velocity of movement in relation to age and muscle morphology. *J. Appl. Physiol.* **46**, 451–456

20. Bassey, E.J. and Harries, U.J. (1990) Torque-velocity relationships for the knee extensors in women in their 3rd and 7th decades. *Eur. J. Appl. Physiol.* **60**, 187–190

21. Larsson, L.K. (1978) Isometric and dynamic endurance as a function of age and skeletal muscle characteristics. *Acta Physiol. Scand.* **104**, 129–136

22. Kohrt, W.M., Snead, D.B., Slatopolsky, E. and Birge, J.S. (1995) Additive effects of weight-bearing exercise and estrogen on bone mineral density in older women. *J. Bone Min. Res.* **10**, 1303–1311

23. Bosco, C. and Komi, P.V. (1980) Influence of aging on the mechanical behaviour of leg extensor muscles. *Eur. J. Appl. Physiol.* **45**, 209–219

24. Bassey, E.J., Fiatarone, M.A., O'Neill, E.F., Kelley, M., Evans, W.J. and Lipsitz, L.A. (1992) Leg extensor power and functional performance in very old men and women. *Clin. Sci.* **82**, 321–327

24a. Fiatarone, M., O'Neill, E., Ryan, N., Clements, K., Solares, G., Nelson, M., Roberts, S., Keyahias, J., Lipsitz, L. and Evans, W. (1994) Exercise training and nutritional supplementation for physical frailty in very elderly people. *New Eng. J. Med.* **330**, 1770–1775

24b. Skelton, D.A. and McLaughlin, A.W. (1996) Training functional ability in old age. *Physiotherapy* **82**, 159–167

25. Pollock, M.L., Foster, C., Knapp, D., Rod, J.L. and Schmidt, D.H. (1987) Effect of age and training on aerobic capacity and body composition of master athletes. *J. Appl. Physiol.* **62(127)**, 725–731

26. Kasch, F.W., Boyer, J.L., Van Camp, S.P., Verity, L.S. and Wallace, J.P. (1993) Effect of exercise on cardiovascular aging. *Age Aging* **22**, 5–10

27. Seals, D.R., Hagberg, J.M., Spina, R.J., Rogers, M.A., Schechtman, K.B. and Ehsani, A.A. (1994) Enhanced left ventricular performance in endurance trained older men. *Circulation* **89(1)**, 198–205

28. Makrides, L., Heigenhauser, G.J.F. and Jones, N. (1990) High intensity endurance training in 20- to 30- and 60- to 70- yr-old healthy men. *J. Appl. Physiol.* **69**, 1792–1798

29. Suominen, H., Heikkinen, E., Liesen, H., Michel, D. and Hollman, W. (1977) Effects of 8 weeks endurance training on skeletal muscle metabolism in 56-70 year old sedentary men. *Eur. J. Appl. Physiol.* **37(53)**, 173–180

30. Allen, W.K., Seals, D.R., Hurley, B.F., Ehsani, A.A. and Hagberg, J.M. (1985) Lactate threshold and distance-running peformance in young and older endurance athletes. *J. Appl. Physiol.* **58**, 1281–1284

31. Belman, M.J. and Gasser, G.A. (1991) Exercise training below and above the lactate threshold in the elderly. *Med. Sci. Sport Exercise* **23**, 562–568

32. Rodeheffer, R.J., Gerstenblith, G., Becker, L.C., Fleg, J.L., Weisfeldt, M.L. and Lakatta, E.G. (1984) Exercise cardiac output is maintained with advancing age in healthy human subjects. *Circulation* **69**, 203–213

33. Yong, L.C., Kuller, L.H., Rutan, G. and Bunker, C. (1993) Longitudinal study of blood pressure: changes and determinants from adolescence to middle age C The Dormont High School follow-up study, (1957–1963) to (1989–1990). *Am. J. Epidemiol.* **138(11)**, 973–983

34. Johnson, B.D. and Dempsey, J.A. (1991) Demand versus capacity in the aging pulmonary system. *Exercise Sports Sci. Rev.* **19**, 171

Glossary

Acclimation/acclimatization
Acclimation is generally defined as adaptation to a single environmental stress (e.g. heat), whereas acclimatization is the adaptive response to a variety of such factors.

Actomyosin complex
A complex of actin and myosin which associates and dissociates during muscle contraction and relaxation. (See also **Cross-bridge cycle.**)

Atrial natriuretic factor (ANF)
Peptide synthesized in and released from the atria of the heart in response to atrial distension. Proposed to play a major role in regulating the body's sodium levels during exercise.

Atropine
An anti-muscarinic, competitive antagonist of acetylcholine. It therefore leads to tachycardia by inhibiting vagal influences at the sino-atrial node of the heart.

Baroreceptor
Specialized stretch-receptor nerve endings which respond to variations in local vascular pressure; important in the control of blood pressure. (See also **Valsalva manoeuvre.**)

Bezold-Jarisch reflex
Coronary chemoreflex that results in reduction of heart rate and arterial blood pressure when mediators are introduced into the coronary circulation.

Bradycardia
Slowing of the heart rate. (See also **Valsalva manoeuvre.**)

Catecholamine
Family of phenolic compounds comprising hormones and neurotransmitters, including adrenaline, noradrenaline, dopamine etc. Adrenaline is one of the most important factors in controlling substrate utilization during exercise. Endurance training leads to lower circulating levels of catecholamines which, combined with other factors, leads to increased lipid oxidation during exercise.

Concentric contraction
Shortening of muscle in response to activation, resulting in power being generated and work being done by the muscle.

Contraction
(See **Concentric contraction, Eccentric contraction, Isometric contraction.**)

Contraction–relaxation cycle
(See **Cross-bridge cycle.**)

Cross-bridge cycle
Cycle of association of actin and myosin to form the actomyosin complex, and of dissociation of the complex into its constituent molecules during muscle contraction and relaxation. The products of ATP hydrolysis (ADP and P_i) bind to myosin which then attaches to an actin molecule in a reversible process that leads to muscle stiffness but does not generate force. The release of P_i from the actomyosin complex is thought to initiate changes which result in force generation. ADP is then released from the myosin head of the actomyosin complex, and the complex is able to bind ATP. Binding of ATP to the myosin head weakens the actin–myosin bonds and causes the release of myosin and thence muscle relaxation.

Cross-over point
The power output at which energy from carbohydrate-derived fuels is equal to the energy from lipids. Further increases of power above this point lead to an increase in carbohydrate utilization and decrease in lipid utilization.

Dynamic exercise
A form of muscular activity associated with changes of muscle length.

Dyspnoea
Difficulty in breathing, breathlessness.

Eccentric contraction
Lengthening of an activated muscle as a result of an external force which exceeds that generated by the activation (e.g. lowering a weight, walking downstairs).

Embden–Meyerhof pathway
The glycolytic pathway by which intramuscular stores of glycogen granules can be broken down to form pyruvate.

Energy charge (EC) ratio
Measurement of the degree of phosphorylation of adenine nucleotides. $EC = ([ATP] \cdot 0.5[ADP]) \cdot ([ATP] + [ADP] + [AMP])^{-1}$. A high EC ratio indicates high turnover of ATP.

Fast-twitch (type-II) muscle fibre
Type of striated muscle fibre which contracts and relaxes quickly. There are two types of fast muscle fibre: fast fatigue-resistant or type-IIA, and fast fatiguable or type-IIB (or type-IIX). Type-IIB fibres generate ATP predominantly by anaerobic glycolysis of glucose and glycogen, whereas the type-IIA fibre is capable of more-aerobic energy transfer.

Fatigue
A complex term that reflects impaired function as a result of prior activity. In a muscle, for example, it is manifest as a reduced ability to generate force. Fatigue typically, therefore, results in failure to sustain physical activity.

Fick equation
The quantitative relationship characterizing oxygen utilization ($\dot{V}_{O_2}$) during exercise. Muscle $\dot{V}_{O_2}$, for example, may be determined from its blood flow multiplied by the difference in arterial oxygen concentration and mixed muscle venous oxygen concentration.

Force–velocity relationship
Relationship between the force generated by a muscle and the velocity of lengthening or shortening of that muscle. As the velocity of shortening decreases, the force that is generated increases; during lengthening, the force increases further.

Glycogenolysis
The process of breaking down glycogen in energy-transfer reactions.

Glycolysis
The process of breaking down glucose in energy-transfer reactions.

Isokinetic strength
The magnitude of the force generated during a contraction with muscle length changing at a constant rate.

Isometric strength
The magnitude of the force generated by a muscle when muscle length remains constant.

Isometric contraction
An activation of muscle(s) in which muscle length remains constant. The joint(s) spanned by the muscle are consequently held in a fixed position. (See also **Static exercise**.)

Hyperoxia
An increase in blood or tissue O_2 partial pressure. This results either from breathing gases with increased O_2 concentration or from alveolar ventilation increasing proportionally more than the body's O_2 utilization rate.

Hyperpnoea
Occurs when minute ventilation is increased.

Hypertension
A persistently elevated level of arterial blood pressure above the normal range for the age group of the individual. May be caused by several different mechanisms.

Hypertrophy
An increase in the size of a muscle cell as a result of exposure to increased work load.

Hyperventilation
A condition associated with reduced alveolar and arterial blood CO_2 partial pressures. Results from alveolar ventilation increasing proportionally more than the body's CO_2 output rate.

Hypocapnia
A result of hyperventilation, i.e. a decrease in the CO_2 partial pressure in the blood or lungs. Caused by alveolar ventilation being increased proportionally more than the body's CO_2 output.

Hypotension
Arterial blood pressure below the normal range (often occurs as the result of haemorrhage) or profound vasodilatation.

Hypoxaemia
A reduction in the blood (usually arterial) O_2 partial pressure. Not necessarily associated with reduced blood O_2 content. For example, a subject sojourning at high altitude may have an arterial blood O_2 content that is normal or even high (as a result of increased haematocrit) but, as the O_2 partial pressure is low, the subject is hypoxaemic.

Hypoxia
A reduction in the availability of O_2 to the peripheral tissues. This may be due to a low arterial blood O_2 partial pressure (hypoxic hypoxia), low tissue blood flow (stagnant hypoxia), reduced O_2-carrying capacity of the blood (anaemic hypoxia) or cellular metabolic poisoning (histotoxic hypoxia).

Lactate shuttle
Mechanism by which lactate formed in some cells and tissues can be oxidized in other cells and tissues, in addition to its role in hepatic gluconeogenesis.

Lactate threshold
The threshold work rate (or oxygen uptake) at which sustained metabolic (usually lactic) acidaemia occurs. Can be used to define the point at which exercise intensity switches from being moderate to heavy. Usually associated with increased ventilatory drive.

Lactic acid
By-product of anaerobic metabolism during exercise. Whereas lactic acid was once viewed as a 'dead-end metabolite' contributing to the oxygen deficit and to fatigue, its formation, exchange and utilization are now considered also to be an important means of distributing carbohydrate energy sources.

Length–tension relationship
Relationship between the length of a muscle and the tension which it can generate.

Maximal oxygen utilization
($\dot{V}_{O_2max}$)
Maximum ability of the body to utilize O_2 during exercise; often used as a frame of reference for characterizing exercise intensity. $\dot{V}_{O_2max}$ is usually taken to indicate maximum aerobic capacity with respect to physical training, etc. The maximum rate of O_2 utilization is well correlated with the volume of mitochondria within a unit volume of muscle fibre.

Maximum voluntary contraction (MVC)
The maximum force that can be obtained by voluntary contraction of a muscle (as opposed to that obtained by electrical stimulation of the muscle).

Metabolic acidaemia
The increase in blood and tissue acidity (i.e. decrease in pH) typically associated with increased production rates of fixed acids, such as lactic acid.

Motor unit
A motor neuron in the spinal cord and all of the muscle fibres that it innervates via its axonal branches.

M wave
The muscle action potential recorded during electrical stimulation of the motor nerve. The M wave is the sum of the action potentials of individual fibres in the part of the muscle near the electrode.

Needle-biopsy technique
Widely used, invasive technique for taking samples of muscle. Samples are rapidly frozen and later sectioned and stained. The technique allows investigators to assess the proportion and distribution of various fibre types in muscle, cross-sectional area of fibres, concentration of enzymes, and mitochondrial and capillary density.

Neuromuscular junction (NMJ)
The cleft or junction between the muscle fibre membrane and the axonal terminal of its innervating motor nerve.

Oral rehydration solution (ORS)
Commercially produced rehydration solutions, usually rich in sodium (90 mM) and potassium (25 mM), as recommended by the World Health Organization in the treatment of diarrhoea-induced dehydration.

Oxygen deficit
The O_2 equivalent of the energy that is utilized during exercise and not accounted for by the measured O_2 uptake from the atmosphere. Includes the depletion of: O_2 stored in haemoglobin and myoglobin and dissolved in blood and tissue fluids; depletion of intramuscular high-energy phosphate stores; and also, at higher work rates, lactate production.

Phosphocreatine
A high-energy phosphate which is degraded during the muscle contraction–relaxation cycle to form the breakdown products creatine and inorganic phosphate to resynthesize ATP.

Phosphocreatine circuit ('energy shuttle')
The mechanism by which phosphocreatine degraded during muscle contraction and relaxation is replenished. ATP (synthesized from ADP in mitochondria via oxidative rephosphorylation) is transported across the mitochondrial membrane to the sarcoplasm in a reaction catalysed by mitochondrial creatine kinase. Once in the sarcoplasm, the ATP combines with creatine to replenish the pool of phosphocreatine.

Plasticity
Change in the contractile and metabolic properties of muscle as a result of the effects of exercise, changes in muscle temperature, etc. Chronic changes in muscle properties may be the result of (e.g.) training, ageing and immobilization. Acute changes may be the consequence of exercise-induced potentiation and fatigue, and changes in muscle temperature.

Propranolol
A nonselective β-adrenergic blocking agent whose effect is most prominent under conditions of high sympathetic stimulation, e.g. exercise. Dominant effect is reduced heart rate and cardiac output. However, as it can also block bronchial smooth muscle β-receptors, its use is discouraged for subjects with bronchial asthma.

Respiratory gas exchange ratio (RER)
Ratio of the rate of CO_2 generation to the rate of O_2 utilization measured at the lung. The RER differs from the Respiratory quotient (see Respiratory quotient) as it also accounts for transient effects of CO_2 washout or storage resulting from hyperventilation or hypoventilation, respectively.

Respiratory quotient (RQ)
The molecular ratio of CO_2 generated to O_2 consumed in metabolic reactions at the tissues. Can be used to estimate the mix of carbohydrate and lipid oxidized under various circumstances.

Slow-twitch (type-I) muscle fibre
Type of striated muscle fibre which contracts and relaxes slowly. Slow muscle fibres generate ATP predominantly by aerobic catabolism of carbohydrate and fats.

Splanchnic circulation
Blood circulation to the liver, gastrointestinal tract, pancreas and spleen: highly compliant.

Static exercise
Exercise involving isometric muscular contraction. Characteristic, for example, of the motion-free phase of the activity of two equally matched arm wrestlers. (See also **Isometric contraction.**)

Stretch–shortening cycle
Combination of different types of contraction: an eccentric contraction (where muscle is activated and lengthened) immediately precedes a concentric (muscle is activated and shortened), power-generating phase.

Sympatholytics
Drugs that block the response to sympathetic nerve stimulation or to sympathomimetic drugs.

T-tubules
T-tubules (or transverse tubules) are invaginations of the plasma membrane that surrounds the muscle fibre. Depolarization of the T-tubule membrane activates the calcium pump in the sarcoplasmic reticulum. The release of calcium ions into the cytosol leads to muscle contraction.

Valsalva manoeuvre
Attempt to forcefully exhale against an occluded airway. The increased intrathoracic pressure results in a complex cluster of cardiovascular reflex mechanisms reflective of the normalcy of autonomic system control. Normally, the initial increase in arterial blood pressure impedes venous return which then reduces the blood pressure, causing peripheral vasoconstriction. The increased venous return that results from the sudden fall of intrathoracic pressure at the end of the manoeuvre causes cardiac output to increase, which (with the residual peripheral vasoconstriction) leads to a large and transient increase in blood pressure, in turn, triggering a bradycardia.

Vasoconstriction
Narrowing of the lumen of a blood vessel often, but not exclusively, as a result of increased sympathetic tone.

Vasodilatation
Widening of the lumen of a blood vessel commonly resulting from the effects of increases in the local concentration of metabolites or reductions of sympathetic activity.

$\dot{V}_{O_2 max}$
(See **Maximum oxygen utilization.**)

Index

acetyl-CoA, 69
actin, 5, 13, 14, 29
action potential waveform, 7, 8, 9
actomyosin ATPase, 17, 29, 36, 51
ADP, 10, 30, 31, 35, 36
adrenaline, 71
aerobic metabolism, 31–33, 40–45, 53
aging, 43–44, 155–169
altitude, 111–112
alveolar gas exchange, 61, 64, 137–138
anaerobic glycolysis, 52
antidiuretic hormone, 84
arterial baroreceptor, 103–104
arterial occlusion, 102
ATP, 4, 10, 17, 18, 29–38, 41, 42, 51–54, 139, 140
ATPase, 17, 29, 36, 51
autonomic activity, 103

baroreceptor, 103–104
blood flow
  in cerebrum, 107–109
  in muscle, 67, 105, 119, 167
  pulmonary, 149–151
  shunting, 66
  in skin, 82
blood–gas partial pressure, 137
blood pressure, 97, 98, 99, 100, 102, 103, 109, 167
body temperature, 78, 79, 81, 83, 91
breathing, 144, 146

Ca²⁺, 6, 9, 29, 69
calorimetry, 61–63
capillary network, 39, 149, 150
capillary-to-muscle fibre ratio, 166
carbohydrate, 61–76, 86, 88
cardiac output, 105–109, 145, 167
cardiopulmonary afferent nerve, 104
cardiovascular
  regulation, 95–105
  system, 38, 166
  transfer, 129–132
central command, 95–96, 104

cerebral blood flow, 107–109
chronic disease, 155
circulation, 95–105
Cl⁻, 82
CO₂
  capacitative storage, 139, 142
  exchange, 132–134
  output, 133, 134, 137–140, 142
contraction, 2, 14–16, 21, 29, 36, 41, 161
contraction–relaxation cycle, 29
contraction velocity (see shortening velocity)
cost of ventilation (see also ventilation), 143–146
creatine kinase, 30, 31, 36, 52
creatine supplementation, 55
cross bridge, 4, 5, 9, 14, 22
cross-over concept, 61–76

dead-space volume, 143
dehydration, 81, 89–90
development, 43–44
drinking, 88–89
dynamic
  exercise, 98, 100–102, 105
  force, 9–10, 49–50

electrolyte
  balance, 84–85, 87, 88, 89–90, 91
  loss, 82–83
electron-transport chain, 117
endurance, 161–162, 165–166
endurance training (see training)
energy expenditure, 29, 40
environmental temperature, 77, 78, 81
evaporation, 80
excretion, 84, 85
exercise
  duration, 53–54
  high-intensity, 49–60, 125–128, 131, 132
  at high temperature, 77–78, 81
  hyperpnoea, 145, 146
  intensity, 53–54
  maximum, 36, 51–55, 109–111, 143, 146, 148, 166

model, 49–51
onset, 97–98
recovery from, 56–57
sub-maximal, 62, 88, 119
exercise pressor reflex, 96–97, 100, 104

fast muscle fibre, 16–25, 37, 58, 59, 69, 125, 159, 163
fatigue, 1–12, 85, 127, 128, 147, 148, 161–162, 164
Fick equation, 41, 119, 120, 128
fluid
  balance, 87, 88, 89–90
  loss, 78–80
  replacement, 85–89
force–frequency curve, 20
force–velocity relationship, 9, 10, 14
free fatty acid, 35, 64, 65, 66, 67, 69

gas exchange, 61, 64, 137–138
glucose, 33–34, 64–67, 86, 88
GLUT-4, 68
glycogen, 33–34, 35, 54, 57, 59, 63, 64, 71, 111
glycogenolysis, 52–53, 68–69
glycolysis, 33–35, 51, 52–53, 62, 68–69
growth hormone, 162

heart rate, 97, 98, 99, 110
heat
  acclimation, 90–92
  production, 79, 80
  stress, 78
high-energy phosphate, 119, 123, 127
high-intensity exercise, 49–60, 125–128, 131, 132
hyperpnoea, 145, 146
hypertrophy, 160
hypothalamus, 80–81, 95
hypoxaemia, 150
hypoxia, 111, 112, 124, 130

inorganic phosphate, 4–6, 29, 30, 32, 41
intracellular pH, 6
isokinetic strength, 156, 158
isometric
  exercise, 97–100
  force, 10, 14, 15, 24, 49
  strength, 22, 156

K^+, 7, 82, 83, 85, 90
kidney, 84

lactate, 52, 57, 62, 66, 71, 110, 111, 112, 125
lactate shuttle, 72, 73
lactate threshold, 73–74, 123–125, 127, 128, 139, 148, 166
lactic acidosis, 67, 71, 124

limitation
  of maximum oxygen uptake, 118
  of oxygen utilization, 128–132
  of ventilation, 147–152
lipid, 61–76
lung volume, 146

malonyl-CoA, 69
marathon running, 62, 63, 79, 80
maximum
  contraction, 2, 21, 36
  exercise, 36, 51–55, 109–111, 143, 146, 148, 166
  heart rate, 165, 166
  oxygen uptake (see also $\dot{V}_{O_2max}$), 110, 130, 131, 164, 165
mechanoreceptor, 103
metabolic acidosis (see also lactic acidosis), 141, 143, 146
mitochondrial
  free fatty acid uptake, 69
  respiration, 32, 34–36, 37
mitochondrion, 31, 39, 41, 43, 69, 166
motor unit, 18
muscle
  afferent nerves, 103
  blood flow, 67, 105, 119, 167
  contraction, 2, 14–16, 21, 29, 36, 41, 161
  endurance, 161–162, 165–166
  fatigue, 1–12
  fibre isoform, 16–25, 37, 58, 59, 69, 125, 159, 160, 163
  fibre recruitment, 69–70
  length–tension relationship, 13–14
  mass reduction, 159
  metabolite, 3–6, 50–51
  mitochondrial reticulum, 70
  oxygen, 38, 119–120, 123, 127
  performance, 26
  relaxation, 9
  single-fibre metabolism, 57–59
  strength change, 155, 156–163
  transfer, 132
muscle-biopsy techniques, 3, 6, 50–51, 61, 159
myocardial ischaemia, 164
myosin, 5, 13, 14, 29
mysosin heavy chain, 16, 17, 18, 23

Na^+, 7, 82, 83, 85, 86, 88, 89, 91
neural control of circulation, 95–105
neuromuscular junction, 3, 6
nuclear magnetic resonance spectroscopy, 3, 4, 6, 41, 123, 128

onset of exercise, 97–98
osmolality, 84, 89

oxidative metabolism, 31–33, 40–45, 53
oxygen
  availability, 124
  consumption, 40, 41, 42, 117
  delivery, 38, 39, 66, 111
  exchange, 117–132
  uptake, 43, 101, 110, 117, 118, 120–123, 125–128, 130, 131, 137–140, 142, 164, 165

perfusion limitation, 148–152
phosphocreatine, 4, 6, 18, 29–32, 36, 41, 42, 51–52, 53–54, 57, 59
phosphocreatine/creatine shuttle, 55
physical performance, 22–25
plasma osmolality, 84, 89
plasticity, 18, 25
power output, 20, 21, 23–24, 40, 64, 69, 79, 163–168
power-to-weight ratio, 165
power–velocity relationship, 24
pulmonary blood flow, 149–151
pulmonary CO_2 exchange, 132–134
pulmonary CO_2 output, 133, 134, 137–140, 142
pulmonary oxygen uptake, 117, 118, 120–123, 125–128, 131, 137–140, 142
pulmonary O_2 transfer, 128–129
pulmonary ventilation (see also ventilation), 110

recovery from exercise, 56–57
red blood cell, 39–40
rehydration, 89–90
respiratory-chain complex, 36
respiratory gas exchange, 61, 64, 137–138
respiratory gas exchange ratio, 61, 62, 70–71, 74, 121, 131, 133, 139, 140, 142
respiratory muscle, 145
respiratory quotient, 62, 74, 121, 133, 139

sarcolemma, 160
sarcomere, 21, 22
senescence, 43–44, 155–169
shortening velocity, 10, 14, 17, 23, 24
shunting of blood flow, 66

single-fibre metabolism, 57–59
skin blood flow, 82
slow muscle fibre, 16–25, 37, 58, 59, 69, 125, 159
solute balance (see electrolyte balance)
speed of muscular contraction, 161
static exercise (see isometric exercise)
sub-maximal exercise, 62, 88, 119
substrate
  delivery, 66–67
  oxidation, 64–70
  utilization, 62
sweating, 79–83, 90, 91
sympathetic nervous system, 70, 71, 72, 95, 99, 105

temperature
  acclimation, 90–92
  body, 78, 79, 81, 83, 91
  environmental, 77, 78, 81
  exercise at high temperature, 77–78, 81
thermoregulation, 79, 80, 83, 87
thermoregulatory vasodilatation, 107
thirst, 81, 85
tissue hypoxia, 124
training
  and cardiovascular function, 111, 112–113
  and lipid stores, 70–72
  and metabolism, 44–45, 56
  in old age, 158–159, 160, 164, 165, 167
  and the pulmonary system, 150, 151
tricarboxylic acid cycle, 34, 35
T-tubule, 7

vasoconstriction, 82, 107
vasodilatation, 107
ventilation
  cost, 143–146
  and exercise, 110, 111, 137–143
  limitation, 147–152
V_{max} (see also shortening velocity), 17, 22, 24, 58
$\dot{V}_{O_2max}$, 33, 34, 37–44, 53, 64–66, 108–111, 113

water balance (see also fluid balance), 84–85